Seeking Sustainable Development on a Level Playing Field

Humans have exploited a huge diversity of materials throughout history. Today's conflict between rising demands and dwindling resources raises searching questions about how optimally to meet humanity's needs efficiently and safely, challenging common assumptions. Plastics support many facets of modern life yet raise associated problems, whilst 'natural' materials may be far from benign when inputs extending their longevity are considered.

PVC (polyvinyl chloride) is a plastic with adaptable, durable and other properties used in diverse construction, medical, information technology, domestic and many other applications besides. However, PVC has faced significant NGO pressure relating to its chlorine content and the range of additives conferring desirable properties. Yet, unlike organochlorine pesticides, PVC plastic is inert and recyclable after providing long service life. This book is not 'pro-PVC' but draws on lessons learned from how the PVC value chain, particularly across Europe, has engaged with problems and made further progress under voluntary commitments to sustainable development.

This book advocates a 'level playing field' of common sustainability principles for assessment of the benefits and risks of the use of all materials in the context of their incorporation within whole product life cycles, from raw material extraction to beyond end-of-life. The use of every material raises specific challenges, but also shares common problems arising from society's legacy of wasteful, linear resource use.

Activities surrounding the PVC value chain have generated novel ideas, assessment techniques and reconsideration of regulatory approaches relevant to sustainability assessment of the use of all materials in the context of whole product life cycles on a common 'level playing field', which best supports the meeting of the diversity of human needs in the safest and most efficient manner.

This book is aimed at industry, regulatory and NGO audiences and influence on wider media.

Seeking Sustainable Development on a Level Playing Field

A PVC Case Study

Mark Everard

CRC Press
Taylor & Francis Group
Boca Raton London New York

CRC Press is an imprint of the
Taylor & Francis Group, an **informa** business

Shutterstock cover image https://www.shutterstock.com/image-photo/technology-hand-holding-environment-icons-over-1913579515

First edition published 2024
by CRC Press
2385 NW Executive Center Drive, Suite 320, Boca Raton FL 33431

and by CRC Press
4 Park Square, Milton Park, Abingdon, Oxon, OX14 4RN

CRC Press is an imprint of Taylor & Francis Group, LLC

Library of Congress Cataloging-in-Publication Data
Names: Everard, Mark, author.
Title: Seeking sustainable development on a level playing field : a PVC case study / Mark Everard.
Description: First edition. | Boca Raton, FL : CRC Press, 2024. | Includes bibliographical references and index. | Summary: "PVC (polyvinyl chloride) is a plastic with adaptable, durable and other properties used in diverse construction, medical, information technology. Yet, unlike organochlorine pesticides, PVC plastic is inert and recyclable after providing long service life. This book is not 'pro-PVC', but draws on lessons learned from how the PVC value chain, has engaged with problems and made further progress under voluntary commitments to sustainable development. The book advocates a 'level playing field' of common sustainability principles for assessment of the benefits and risks of the use of all materials in the context of whole product life cycles"–Provided by publisher.
Identifiers: LCCN 2023030366 (print) | LCCN 2023030367 (ebook) | ISBN 9781032592770 (hardback) | ISBN 9781032590196 (paperback) | ISBN 9781003453949 (ebook)
Subjects: LCSH: Polyvinyl chloride industry–Environmental aspects. | Plastics industry and trade–Environmental aspects. | Recycling (Waste, etc.) | Sustainable development.
Classification: LCC HD9662.P62 E94 2024 (print) | LCC HD9662.P62 (ebook) | DDC 381/.456684236–dc23/eng/20230920
LC record available at https://lccn.loc.gov/2023030366
LC ebook record available at https://lccn.loc.gov/2023030367

ISBN: 9781032592770 (hbk)
ISBN: 9781032590196 (pbk)
ISBN: 9781003453949 (ebk)

DOI: 10.1201/9781003453949

Typeset in Times
by codeMantra

Contents

Foreword		vii
Acknowledgements		ix
About the Author		x

1 Introduction — **1**

2 Living Chemistry — **5**
2.1 Organisms and Chemicals — 5
2.2 Human Ingenuity — 8
2.3 A History of Humanity through Chemical and Material Use — 9
2.4 'Better Materials' to Meet Human Needs — 12
Notes — 13

3 Problematic Chemistry and Sustainable Development — **14**
3.1 Nature's Chemical Armoury — 14
3.2 Human Uses of Chemicals — 17
3.3 Human Abuses of Chemicals — 18
3.4 Societal Responses to Chemicals of Potential Concern — 22
3.5 Serving Human Needs — 28
Notes — 30

4 PVC: The Good, the Bad and the Prejudiced — **33**
4.1 What is PVC? — 33
4.2 The Dark Side of PVC — 37
4.3 Pressure on the European and UK PVC Sector and the Early Responses — 41
4.4 Industry Engagement with Sustainable Development — 43
4.5 Breaking through the Prejudgements — 52
Notes — 55

**5 Voluntary Sustainability Commitments
   of the European PVC Sector**   **58**
   5.1   Vinyl2010   59
   5.2   VinylPlus® to 2020   59
   5.3   Addressing the UN Sustainable Development Goals   60
   5.4   VinylPlus to 2030   63
   5.5   Evolving VinylPlus Accreditation Schemes   64
   5.6   Additive Sustainability Footprint (ASF)   67
   5.7   Progress with VinylPlus Targets   70
   5.8   Novel PVC Products and Life Cycles Driven
         by Sustainable Development Challenges   70
   5.9   Elsewhere in the World   83
   Notes   86

6 A Level Playing Field?   **89**
   6.1   Wider Comparative Assessment Using the
         Five TNS Sustainability Challenges for PVC   91
   6.2   A Level Playing Field: ASF   96
   6.3   Necessary Evolutions in Regulation,
         Assumptions and Common Understanding   107
   Notes   108

**7 Sustainability and the Purpose of Business
   and Regulation**   **112**
   7.1   The Business of Business   112
   7.2   The Business of Regulation   116
   7.3   The Business of Strategic Regulation in an
         Imperfect World   118
   7.4   The Business of Regulation in a Globalised World   122
   7.5   A Global Challenge   123
   7.6   The Business of Leadership   123
   Notes   124

8 Epilogue   **126**

Index   129

Foreword

Foreword by Jonathon Porritt
Founder Director of Forum for the Future, author and campaigner

Sesame Street's Kermit the Frog is a global icon for many reasons, including his oft-quoted quip that 'it's not that easy being green'. But as both Mark Everard and I know only too well, Kermit's problems are nothing in comparison to being a devotee of sustainable development!

It was in 1987, with the publication of the Brundtland Report, 'Our Common Future', that the concept of sustainable development began to get traction. Defined as 'development that meets the needs of the present without compromising the ability of future generations to meet their own needs', it forced both the development movement and the environment movement to take off their blinkers and start looking at things in a much more integrated and holistic way. For me, it remains the most important and radical 'big idea' in the world today.

But it's not an easy idea, uncomfortably challenging people's prejudices, demanding analytical acumen and only making a real difference when the principles that lie behind it are turned into hard-edged practice, in real-world situations.

That's what I love about *Seeking Sustainable Development on a Level Playing Field: A PVC Case Study*. It works at many different levels: historical and philosophical; applied and theoretical; educational and personal. Mark Everard is one of the 'great contextualisers' in the world of sustainable development, constantly stretching the scope of inquiry, always acting as an eloquent destroyer of narrow, blinkered thinking.

We often had to remind ourselves of the importance of these principles when we worked together on an initiative called the *PVC Coordination Group* between 1998 and 2002. At that time, my organisation, Forum for the Future, was the host in the UK for The Natural Step, a Swedish conceptual process and organisation that applies rigorous System Conditions to help people understand the essence of sustainable development.

Our Coordination Group took over from something called the *PVC Retailer Working Group*, which was facilitated by Greenpeace. But when the research done by that Group did not confirm Greenpeace's position at that time (Greenpeace had campaigned against the industrial production of chlorine for decades), they reverted to a more familiar (and scientifically unsound) position of pressing ahead with their campaign *regardless*, enthusiastically demonising PVC as 'an inherently wicked material' for which any number of alternative materials would provide better substitutes.

That just seemed bonkers to us. But it was an effective campaign, which many businesses rapidly bought into as they set out to demonstrate their green credentials by phasing out their use of PVC.

In Chapter 4, Mark tells of an encounter with a leading sportswear manufacturer in Europe seeking his approval (unsuccessfully!) for its decision to get out of PVC. I had exactly the same battle with a leading UK retailer who could see the way the wind was blowing in terms of public opinion and wanted to be 'the first' to phase out PVC in all its packaging materials, regardless of the science. As Mark says: 'I'm not pro-PVC: I'm simply anti-stupidity'.

Hence his appeal for a genuinely level playing field when different materials are being assessed. This rarely happens. What Mark calls 'prejudgements' often get in the way.

However, the benefits of sticking to first principles, and engaging constructively and transparently with companies seeking to make a genuine difference, are there for all to see in this book. And a lot of credit here goes to the 'pioneers' in the PVC industry itself, back in the 1990s, particularly Norsk Hydro, that not only had to embrace often very unwelcome challenges in those early days, but has had the tenacity to stick with this 'agenda for transformation' since then.

The evolution of that first Stakeholder Forum into the ECVM's Charter, and then into Vinyl2010, the VinylPlus programme to 2020 and now the VinylPlus Pathways to 2030, speaks volumes. This is an industry that still faces some huge challenges, particularly in terms of 'closing the loop' through far more ambitious recycling schemes on a global basis. But its early embrace of a rigorous Framework for Strategic Sustainable Development, and a readiness to meet those challenges through investment in innovation, needs to be commended – whatever you may or may not think about PVC.

As we all know, the plastics industry as a whole is still very much in the dock today, with its massive disbenefits (ocean plastics, microfibres and so on) continuing to obscure the significant social and economic benefits that a vast range of plastic materials provide us with.

Far from being a 'pariah material', endlessly demonised by those seeking relatively easy targets, PVC still has an important role to play in that global challenge.

Acknowledgements

The work reported in this book has resulted from the vision and labour of many people with whom I have interacted over a quarter of a century. These people are from NGO, business, academic and regulatory backgrounds. Some of them I have clashed with creatively. With many more, I have worked strategically, developing and sharing a vision of the contributions of materials science and use towards a better, sustainable and equitable future. A good number – both opponents and supporters of PVC as well as others in different branches of sustainable development advocacy – have become friends and comrades in a common mission.

As a young idealist, researcher and protestor in the dawning days of the modern 'environment movement' of the mid-1970s, now a half-century ago, I was often told I would grow out of it and accept life as it was. I am thankful to all these people for indulging the childish disbelief that I have never managed to shed!

To allay any accusation to the contrary, I received no funding or coercion to write this book. It was written because I woke up on 9 March 2023 realising that it was needed to inject some strategic sense into something of a moribund and polarised debate about public opinions, assessment and regulation of chemicals. We need sophisticated chemistry to help booming numbers of people in global society to meet their needs with greater efficiency, safety and equity in the face of increasing and now looming existential threats. So, let's all get on with the job of working within our spheres of influence to co-create that mooted sustainable future!

About the Author

Professor Mark Everard is a Visiting Professor at Bournemouth University, as well as Associate Professor of Ecosystem Services at the University of the West of England (UWE Bristol). He also works as a consultant, broadcaster and author. Mark is Vice-President of the Institution of Environmental Sciences (IES), a Fellow of the Linnean Society, an Angling Trust Ambassador, and a science advisor to WildFish (formerly Salmon &Trout Conservation UK), Tiger Water (India), Wiltshire Wildlife Trust and a range of other bodies. Mark has worked with the PVC sector since 1999, at that time as Director of Science with the international NGO The Natural Step (TNS), developing the five TNS Sustainability Challenges for PVC that have since been embodied in revised form as the five key challenges of the VinylPlus voluntary commitment across the European PVC sector. Mark continues to work with the PVC sector amongst other businesses and policy areas, as well as serving in academia and broadcast media, to promote practical progress with society's greatest sustainability challenges. Mark's work with PVC and other materials is part of a wider portfolio of systems and sustainable development research, advocacy and communication (including 40 books and over 130 peer-reviewed scientific papers as well as frequent magazine, TV and radio contributions) on natural resource management particularly in the developing world and on rivers, wetlands and catchment management, and a range of other disciplines.

Introduction

1

Polyvinyl chloride (PVC) is a contentious material. It is, though, more correct to say that PVC is a material with which contention has been associated. In reality, PVC is one amongst a huge variety of materials used by society to meet its varied needs. It is, however, a substance that has received a disproportionately higher level of scrutiny than other bulk chemicals used by society, creating perceptions in some quarters that it is a material apart from all others. A collateral impact of this perception is that these other materials have, by and large, escaped equivalent scrutiny, often benefitting from lazy implicit assumptions that they are automatically more sustainable.

PVC certainly has issues to be addressed. Some relate to its chlorine content, and others to some of the historically used problematic additives imbuing PVC compounds with desirable properties, with other issues related to the persistence of plastic waste where recovery and recycling have not been achieved. Regulatory attention has highlighted some of these issues, and progress has since been made, in many instances on a voluntary basis. However, a variety of studies also suggest that PVC does not differ significantly from other materials, sharing many challenges common to other substances driven from society's unsustainable resource use habits whilst also, like others, having some of its own unique sustainability concerns.

Chapter 2, *Living Chemistry*, explores the role of chemicals in the natural world from the origins of terrestrial life and onwards to the many diverse and sometimes surprising solutions subsequently evolved by plants, animals and microbes, as well as the often-vital interactions between them. The same is true of humans, which interact with chemicals and materials in their environment to meet multiple needs ranging from basic biophysical requirements through to economic activities and innovations that have underpinned many aspects of cultural evolution. Discoveries of the properties of different chemicals and their manipulation to meet a range of needs have been part of the defining features of human progression from prehistory, throughout history and through to the present day. Impressive innovations in chemical use have supported agricultural, industrial, medical, economic and many wider

DOI: 10.1201/9781003453949-1

advancements. The emergence of the 'age of plastics', as one aspect of synthetic chemistry, has been highly significant and now pervades many facets of the modern world.

Chapter 3, *Problematic Chemistry and Sustainable Development*, examines some of the problems created by the use and abuse of chemicals, ultimately driving humanity to recognise the importance of developing on a more sustainable pathway. Human uses of substances of many types can be considered abusive if they contribute to the degradation of supportive ecosystems, or that they more directly undermine the capacities of others to meet their needs. Human exploitation of natural chemistry and the innovation and use of synthetic substances to meet a variety of needs have often formerly had little regard for their potential for these wider environmental and social abuses. Emerging awareness of adverse environmental and human health consequences arising from incautious or large-scale use of a variety of substances drove reactive business and regulatory responses. This emergence of awareness was also a foundation of the modern 'environment movement', leading through to contemporary understandings and rhetoric about sustainable development. Early reactivity can be transformed into a proactive approach informed by understanding and application of the scientific principles underpinning sustainability, better to guide material choice, use and innovation, potentially on a more profitable basis if it better serves the meeting of human needs in safer and more efficient ways.

Chapter 4, *PVC: The Good, the Bad and the Prejudiced*, introduces the polymer polyvinyl chloride (PVC), aspects of its manufacture and the additive substances with which the polymer is compounded as a finished plastic. The chapter explores the positive uses of this plastic, some associated problems, pressure on the PVC industry particularly from the NGO sector and also from regulatory scrutiny and some early responses. Engagement with sustainable development by the PVC industry, particularly in the UK but spreading more widely across Europe, is then reviewed. Criticisms levelled at PVC are explored, some justified and some based on unfounded prejudgements. Proactive progress with sustainability challenges is acknowledged whilst also reflecting on a long road yet to be travelled to achieve the ultimate goal of sustainability, with many challenges common to PVC and other materials alike. The focus of this book is on a level playing field for sustainability assessment relevant to all material types. However, the asymmetric attention directed at PVC has led to proactive engagement with and voluntary commitments to sustainable development by players in the PVC value chain, with associated innovations that are generically informative as case studies for society's uses of other materials.

Chapter 5, *Voluntary Sustainability Commitments of the European PVC Sector*, covers the progressive and evolving engagement of the PVC industry across Europe with sustainable development. This includes, in particular, successive voluntary commitments under the Vinyl2010 and VinylPlus programmes. An evolving suite of VinylPlus accreditation schemes is introduced, outlining their roles in distinguishing PVC-containing products compliant with VinylPlus voluntary commitments to sustainable development. Tools developed to support sustainability assessment and compliance are described, in particular the Additive Sustainability Footprint (ASF) approach developed for assessment of the sustainable use of additives in the context of the life cycles of the products into which these substances are incorporated. The 'line of sight' from underpinning scientific principles informing the System Conditions of The Natural Step (TNS), the Sustainability Challenges emerging from application of these System Conditions to the PVC value chain, and their incorporation into VinylPlus Pathways towards 2030 is emphasised, along with linkages to the UN Sustainable Development Goals. Examples of practical progress against targets under the VinylPlus challenges are outlined. There is also consideration of engagement with sustainable development by the PVC industry in other global regions.

Chapter 6, *A Level Playing Field?* addresses the importance of, and practical steps towards, a common, science-based and systemic approach to sustainability assessment of the use of chemicals, taking account of their contributions to the life cycles of the products into which they are incorporated. Increasing human demands yet dwindling resources stress the importance of a paradigm shift from simplistic hazard-based protocols, progressing to approaches that help recognise how best to meet human needs in the safest and most efficient manner. Comparative assessments using the ASF approach highlight sustainability-relevant issues for different substances in the context of whole product life cycles, including the benefits they confer, associated risks and potential mitigation measures or priorities for innovation. There is a need for paradigmatic change in established assumptions and also in chemical regulation, for which ASF may offer useful insights, if it is to drive innovation and practice towards sustainable outcomes. This systemic perspective can also indicate priority areas for innovation and investment better serving needs and novel markets increasingly shaped by sustainability pressures.

Chapter 7, *Sustainability and the Purpose of Business*, reflects on the changing roles of business throughout history, particularly in the light of the need for the innovation necessary to make progress towards a sustainable future. The capitalist model that now pervades much of the world is society's chosen means to serve human needs by profitably converting natural and human resources into useful products, but profit-generation must not be divorced from social and environmental progress. As sustainability pressures inevitably influence future markets, far-sighted businesses can use strategic understanding of sustainable development to innovate the materials, products and services required to better serve human needs in safer and more efficient ways that are also de-risked and profitable. Regulation also needs to evolve to inform and encourage sustainable progress, recognising and informing incremental steps that lead strategically towards sustainability. It is also necessary for regulatory approaches to not only become more systemically informed, but to address the globalised nature of business.

Chapter 8, *Epilogue*, rounds off this book with concluding reflections concerning progress not only within the PVC sector but how this has ramified out into wider influence on societal choices and assessments of material use informed by a systemic approach, and how the journey of sustainable development is far from finished. The chapter emphasises that the core theme of this book – the need for a level playing field to drive assessment, innovation and investment – is essentially 'material blind', albeit informed by a great deal of work across the PVC value chain over the past quarter-century. In this pivotal time in human history, there is a pressing need to take a more mature, systemic approach to determining how best to meet human needs in the safest and most efficient manner. This book has the purpose of deepening thinking across all societal sectors – chemical and materials industries, procurement departments, other sectors of business, and regulatory and government bodies, amongst others – about what sustainable development means in practical terms with regard to material choices and use, as we all have roles to play in progress towards a sustainable society.

The book's title, *Seeking Sustainable Development on a Level Playing Field: A PVC Case Study*, reflects both how this singular type of plastic has been used and the issues raised that have stimulated engagement of the wider sector with sustainable development. However, the key theme is how progress with thinking about PVC and other substances can inform us about what sustainable use means for the wide variety of materials used by society, and the essential role of a level playing field of assessment to guide chemical choice and innovation, value chain management and regulation, proactively to help society meet its needs in the safest and most efficient manner.

Living Chemistry

2

As a bold start to this book, let's talk about the meaning of life. From a scientific point of view, key characteristics of living matter include that it is responsive, that it grows, has a metabolism, undertakes energy transformation and that it reproduces. These are all active processes. All, in one way or another, entail interacting with surrounding matter, be that living or non-living. Biology is a miasma of complex physics and chemistry, including the genetic molecules encoding the Earth's vast array of life forms and the throughput of chemicals maintaining cellular structure, metabolism and replication.

2.1 ORGANISMS AND CHEMICALS

Although the exact origins of life on Earth are unknown, the earliest organisms appearing on planet Earth around 3.85 billion years ago were autotrophs. An autotroph, in simple terms, is an organism that generates energy and produces its own food. Most likely, these primeval organisms achieved this task by splitting bonds in chemicals within their environment, releasing energy from them to drive metabolic processes and to build the complex chemicals of which they were formed. Autotrophs with these traits still occur widely in many environments today.

Autotrophs produce their own food using water, carbon dioxide or other substances, powering these processes by capturing light or breaking chemical bonds. Because autotrophs produce their own food, they are sometimes called 'producers'.

The earliest life forms lived in an anaerobic or low-oxygen environment. That was all to change around 2.5 billion years ago with the evolution of photosynthesis, harnessing energy directly from solar energy to split water and to forge it with carbon dioxide to generate glucose, a sugar, that was subsequently transformed into a range of organic substances with structural, energetic, enzymatic and other functions. Whilst a major step forwards in terms of

DOI: 10.1201/9781003453949-2

the energetic power of life, it was nonetheless a death knell for many organisms as 'waste' reactive oxygen generated by the process of photosynthesis built up in the atmosphere. This increasing concentration of reactive oxygen drove to extinction a vast proportion of life forms evolved in formerly anaerobic conditions. In passing, as a curio, it is worth reporting that another group of autotrophic fungi known as 'radiotrophs', discovered inside and around the abandoned Chernobyl Nuclear Power Plant in Ukraine, use gamma radiation and the pigment melanin to capture energy for growth.

The proliferation of organisms rich in energy and organic matter enabled the subsequent evolution of heterotrophs: organisms that eat other organisms, either living or as dead organic matter, as a source of energy and chemical constituents. Heterotrophs, also known as 'consumers', may directly consume autotrophs such as algae, fungi or higher plants or, in increasingly complex food webs, may be carnivores (eating other animals), omnivores (eating both plants and animals) or detritivores (eating the dead remains of plants or animals or their excretions). Food webs have amplified in complexity throughout evolutionary history. All creatures today, from the largest whales and trees to the most microscopic soil bacteria, fungi and archaea, play crucial roles in the cycling of energy and matter. Biology maintains physics and chemistry in sustainable and cyclic motion.

Beyond a basic metabolic level, living organisms exploit and manage the substances around them in a bewildering array of complex ways. Termites secrete a glue-like substance to stick together fine earthen particles to construct towers with almost concrete-like consistency, and the physical structures of these towers act as air conditioning vents maintaining cool conditions in underground levels that serve as nurseries for larvae and also 'gardens' where fungi grow on foraged vegetation as the primary food source of the colony. Some termites also manufacture defensive chemical weapons, expelled via a fontanellar gun in the form of a horn-like frontal projection on the head of the soldier caste, used to ward off predators such as ants. Male sticklebacks, small fish of fresh and brackish waters, build nests of aquatic vegetation stuck together with a protein excreted from the kidneys known as spigin, the male fish attracting a sequence of females into the nest to lay eggs that are then fertilised and nurtured to the point where the fry hatch and become free-swimming. Bird nests of sticks, mud, moss, spider webs and other materials are more familiar examples of organisms using materials in their environment to further their biological needs. The nests of crocodiles are built from vegetation that rots down, raising the temperature to incubate the reptile's eggs. Other creatures such as foxes, badgers and rabbits tunnel underground, moving earth for refuge and as nurseries. Wood ants pile the needles of evergreen trees into substantial mounds as a home for the colony. Everywhere, nature is moving, building, gardening and interacting with surrounding chemical and physical structures.

Termites are considered 'ecosystem engineers' as their activities form, destroy, modify or maintain whole habitats in significant ways, creating conditions from which a wide range of other species benefit. They achieve this by maintaining soil health through nutrient cycling, ingestion of organic material and mineral debris, aerating the soil via their burrows and allowing rainwater to permeate, moving around substantial quantities of soil for mound-building, binding soil particles together and fertilising it with their excrement. This creates conditions stimulating the growth of plants, whilst the towers provide hiding places and hunting grounds for other animals. Other ecosystem engineers affecting the chemical cycles, physical structure and biological relationships across whole ecosystems include beavers, which divert water flows and diversify wetlands and meadows exposed though tree felling supporting their dam-building activities. The activities of beavers form new habitats for other aquatic organisms, influencing chemical and energy cycles as well as the hydrology of catchments, and open up more densely shaded areas diversifying landscapes for plants and a wider range wetland and terrestrial organisms. African elephants are another prominent ecosystem engineer, changing their environment and creating habitats for other species through age-old migration trails that sculpt the land with deep grooves and create pools in their footprints, also pushing over trees or removing bark and so opening up landscapes for grassland and the many species that depend upon it. Other ecosystem engineers include corals that create physical structure affecting ocean currents including the upwelling of nutrients, enabling a huge diversity of plants and animals to thrive. Kelp beds too function as underwater forests in shallow seas, affecting temperature and nutrient cycles and providing shelter and food for fish, crustaceans, molluscs, turtles and a wide range of other marine organisms.

At the microscopic level, some of the most complex and intimate chemical exchanges and other interactions between unrelated organisms occur underground in the rhizosphere, comprising the interface of the plant's roots with the soil including a hugely diverse association of microorganisms. Probably all rooted plants growing in soils have a close, possibly vital relationship with a host of microorganisms to which they secrete as much as 40%, and generally at least a fifth, of their net production of sugars and organic acids. This release of energetic and nutritious chemicals may initially seem inefficient, but it demonstrates the magnitude of importance of the many relationships that rooted plants have with fungi, bacteria and other microbes. The intimate interaction between fungal hyphae and rootlets enables the plant to benefit through the actions of fungi that liberate phosphorus, potassium, iron and other minerals from the surrounding soil, resources that are largely inaccessible in the absence of these interactions. Nitrogen is also made available, particularly through the activity of symbiotic bacteria that 'fix' largely inert atmospheric nitrogen into biologically available forms. In addition to nutrient cycling, fungal support also includes disease suppression through the production of antibiotics. A wider biota is associated with the rhizosphere, including a proliferation of nematodes that may have roles in suppression of diseases potentially attacking the plant. The importance of the close microbial association naturally comprising the rhizosphere is exemplified by the relatively poor productivity of crop plants in soils in which microbes and macrofauna are degraded through excessive tilling, pesticide use and, ironically, application of artificial fertilisers that the plants are far less efficient at absorbing. In reality, no plant can thrive alone, as all are part of complex communities of organisms maintaining vital chemical cycles and exchanges. These fungal and other subsoil networks connect plants to others of their own as well as unrelated species sharing the same habitat, including via the exchange of nutrients and other chemicals and the communication of 'warnings' from plants infested with pests enabling non-infected plants to prepare by producing defensive substances.

2.2 HUMAN INGENUITY

As with other species on Earth, the wellbeing of we humans is intimately connected with our chemical interactions with the surrounding environment. But it is our ingenuity in the foresighted use and innovation of substances that marks us apart, bending chemistry, physics and biology to our ends.

From food to clothes, construction materials, tools and natural medicines, humans have used what surrounds them to further their nutrition and wider aspects of health, security and spiritual wellbeing, and significantly also their economic activities.

In many cases, innovation has enabled the extension of natural chemistry to meet particular needs. A classic example here is the innovation of aspirin (acetylsalicylic acid or ASA), a synthetic nonsteroidal anti-inflammatory drug (NSAID) used to treat pain, fever, inflammation and a range of other medical conditions. A precursor of acetylsalicylic acid is salicylic acid, found in the bark of willow trees (willow trees of the genus *Salix* give the drug its common name), that has been used for around 2,500 years for its recognised health benefits. First synthesised in 1853, aspirin is one of the most widely used medications globally.

Human ingenuity goes well beyond exploitation of natural substances for beneficial uses, into synthesis of novel compounds. Some of these, including many families of drugs, may be analogues of natural substances. Others, including many organochlorine substances such as polyvinyl chloride (PVC), are entirely novel. Indeed, much of the modern world rests on humanity's skills in innovating and working with steel, concrete, polymers, ceramics and many more synthetic materials besides.

2.3 A HISTORY OF HUMANITY THROUGH CHEMICAL AND MATERIAL USE

As elaborated in my 2016 book *The Ecosystems Revolution*,[1] the history of cultural evolution can be told as a story of material revolutions. Each of these revolutions or step changes was framed by discovery and progressive societal uptake of materials that were more efficient in enabling people to meet various of their needs.

The most often-told of these waves of material-based innovation is the Stone-Bronze-Iron system, occurring over an extended period from prehistory through to recorded history. This overlapping succession of ages spanned a journey of societal evolution of around six millennia, each step enabled by discovery and manipulation of materials increasingly better suited to the demands of human activities. Each phase was also progressive, entailing piecemeal innovation and sophistication contributing to greater cultural complexity.

The Stone Age (6000 and 2000 BCE) was coincident with evolution of the genus *Homo*. The capabilities of *Homo erectus* to make and develop tools with sharp edges, points or percussive surfaces were a significant factor in the successes of the hominids. Tools of stone and other hard materials constituted a defining feature, spanning three phases – Early Stone Age, Middle Stone Age and Later Stone Age – each leaving behind durable artefacts revealing progressive evolution in the sophistication of tools and the capabilities of their users. Progressions occurred in agriculture and the domestication of animals, advances in social structure, and the exploitation of novel food sources and new environments. Although varieties of stone (flint and chert where a sharp edge was needed; basalt and sandstone for grinding) were the principal materials used, tools were also fashioned from other hard natural substances such as bone, shell and antlers and, later, pottery derived by firing clay. Other diverse aspects of cultural practice and evolution can be assumed, some evidenced by the kinds of tools these people left behind.

The Bronze Age (c3300–1200 BCE) saw the first significant manufactured material: bronze, formed as an alloy of copper and tin. An earlier transitional period was known as the Copper Age when people had learned to smelt copper but not yet to manufacture bronze. Increasing uptake of bronze tools and other items was either due to the spread of the technology or, given the geographical scarcity of tin, trading of bronze artefacts. Significant social evolution observed throughout the Bronze Age included the rise both of Mesopotamian and Egyptian cultures, each of them developing the earliest known writing systems in addition to a range of technologies, social structures and differentiation, and with accompanying advances in science.

Progressive transition into the Iron Age, dated back to 3200 BCE in northern Egypt, saw iron and some steel articles becoming more prevalent for cutting tools and weapons. Understandings of iron metallurgy and purification from oxidised ores spread rapidly across the human world between 1200 and 1000 BCE. Although iron is not significantly harder than bronze, combination with carbon yielded the far harder material of steel. These strong new tools helped drive ahead progressive agricultural practices and artistic styles, and supported proliferation of religious beliefs, written language, literature and historic records, as well as of novel tools, weapons, ornaments, pottery and decorative designs.

Significantly, transition between these ages did not result from the world running out of stone, or indeed bronze though the availability of tin may have been a limiting factor. Rather, it was enabled by an incremental spread of knowledge, technical innovation and, most significantly, recognition of the advantages that exploitation of a new materials better helped people meet their needs. There was considerable overlap between these phases of human material use as techniques were taken up in waves across different parts of the inhabited world.

Other revolutions have been contemporaneous with these material revolutions including, for example, innovations in water management enabling leaps forward in food security, fomenting the establishment of the world's first known settled and differentiated civilisation in the 'Fertile Crescent' of Mesopotamia during the Bronze Age. Indeed, innovations in water management for irrigation, livestock watering, defence, transport and trade, kinetic and subsequently electrical power generation have defined many subsequent human 'revolutions' right up to the present day. This includes the re-plumbing of entire continents, as for example in the case of the massive water diversion and inter-basin transfer schemes implemented in China. Control of the simple chemical H_2O has been foundational in various 'agricultural revolutions' throughout human history and, with it, water-vectored flows of chemical nutrients, progressively enabling transitions from nomadic to transhumance and then sedentary lifestyles, through to the burgeoning of cities to which piped water and the fruits of irrigated landscapes were appropriated from increasingly remote hinterlands.

Notwithstanding chemistry as a science having a long history dating back to Egyptian and Ancient Greek times, one of the most 'revolutionary' periods of chemical innovation occurred during the European Industrial Revolution. The European Industrial Revolution was a period of approximately two centuries of cascading innovations in technology and manufacturing processes, commencing in the English West Midlands and North West from around 1760 and spreading progressively across much of Europe. Transition from animal and human power and biological fuels towards increasing reliance on machines powered by water and energy-dense fuels such as coal went hand-in-hand with novel chemical processes driving forwards metallurgy and innovation of novel materials such as cement. Other of the many progressions in the use of materials included new methods of glass- and paper-making as well as transport systems.

Innovations in chemistry not only changed society through the ongoing roll-out of industrialisation, but in many more ways. Chemical innovation and mechanisation of printing effectively democratised knowledge that had formerly been retained by an educated and privileged minority, fomenting social revolution challenging established class structures. The contributions of advancing chemical knowledge and engineering to medicine were and remain in many ways miraculous, including the innovation and use of novel families of drugs some of which derive from analogues of natural substances. Some more modern drugs integrate with the chemical fabric of genetic matter and immune chemicals.

Natural polymers occur everywhere in nature, and indeed are the substance of life. Chains of organic molecules include our genetic matter, proteins, starches as well as cellulose and other complex carbohydrates. Many of these natural polymers, such as wood, cotton and other plant fibres, as well as wool, feathers and other animal fibres, have been exploited directly by people over centuries. At their most simplistic, this is through splitting of energy-rich carbon-to-carbon and other molecular bonds to release energy, be that as food or fuel. But more innovative and complex uses include the harvesting and/or modification of natural polymers as pharmaceuticals, fabrics, packaging materials such as cellophane or rubber extracted from trees, bitumen extracted from tar deposits and many more applications besides.

The first fully synthetic polymer material, Bakelite, was developed by Leo Hendrik Baekeland in 1907. Bakelite (polyoxybenzylmethylenglycolanhydride) is formed by a condensation reaction between phenol and formaldehyde. Baekeland proceeded to patent the substance and the production process in December 1909. Bakelite has extraordinarily high resistance to electricity, heat and chemical action, and so was particularly well suited for many applications in emerging electrical and automobile industries. The applications of Bakelite were many and varied, including in electrical systems, aeroplane parts and radio during the First World War. And so, the 'Age of Plastic' came into being.

The twentieth century witnessed the innovation of many novel plastic materials. Plastics comprise any of a range of malleable synthetic or semi-synthetic long-chain organic polymer materials of variable rigidity and other properties. These novel plastics found applications in a bewildering diversity of products and technologies ranging from landline and now mobile telephones, music and entertainment technologies including the vinyl record as well as CDs and DVDs, in insulation, packaging, aviation, healthcare, transportation, plumbing, power supply and distribution, toys and sporting equipment, geomembranes, pipes, electronics, renewable energy technologies, and many more construction applications. The huge technological advances over this time period progressed hand-in-hand with the proliferation of different types of plastic, new uses enabled by the inherent properties of these materials including their durability, electrical and thermal insulation, relatively low cost, ease of manufacture, versatility, and imperviousness to water and weathering.

Synthetic plastics as a class of materials are also highly adaptable. The inclusion of additive chemicals such as plasticisers, stabilisers, pigments and impact modifiers modifies their properties to suit an increasingly broad range of uses. These rapid advancements in polymer science were to drive a progressive displacement of traditional materials, such as various metals, leather, natural rubber, wood, stone, glass and ceramics, in many applications across society. One of the most dramatic contributions of synthetic polymer chemistry is observed in healthcare. The readily cleaned and sterilised surfaces of many types of plastic offer safer medical environments, and plastics are widely used for such purposes as durable blood bags, catheters, cannulas and tubing, and durable and inert implants. Plastics have also enabled progress in electronics, such as those behind the development of advanced medical imaging technologies such as X-ray, Magnetic Resonance Imaging (MRI) and Computerised Tomography (CT) as well as ultrasound scanning. Not only do these advances often sound like recent science fiction, but they have supported major medical advances saving many lives (mine included). Synthetic plastics have become fundamental to almost every aspect of contemporary human lives, with the global production of plastics reaching 390 million tonnes in 2022.[2]

The innovation and use of novel silicon, germanium and other semiconductor components, polymers and composites, and many more substances besides have promoted advances in information technology, arms, medical, entertainment and space technologies. Chemical innovations are some of the more pervasive human creations, ranging from development of fertilisers, explosives, propellants, lubricants, drugs, weapons, scents and many more applications. Many are beneficial, though others are potentially hazardous. Chemically enabled advances in human history have been not only momentous, but have delivered and continue to deliver benefits enjoyed by virtually everyone on the planet.

2.4 'BETTER MATERIALS' TO MEET HUMAN NEEDS

For most of our history, the use of cascades of novel materials resulted from discovery or innovation of substances meeting human needs more efficiently. We did not, as we previously observed, exit the Stone Age because we ran out of stones, but because we found better materials to do the jobs that they previously enabled us to do.

However, what we have substantially lacked throughout this history of material exploitation has been a sense of the wider ramifications of our use of substances across whole life cycles from extraction to conversion, use and maintenance, and at and beyond the end of useful life of the products and artefacts into which they have been integrated. With booming human numbers and per capita resource demands, yet on a finite planet of dwindling environmental capacity, this wider perspective is one that we continue to overlook at our peril.

In addressing what 'better materials' means in a world beset by increasing sustainability challenges, we need common frameworks for proactive thinking, decision-making and innovations about how best to meet human needs in the safest and most efficient manner.

NOTES

1 Everard, M. (2016). *The Ecosystems Revolution: Co-creating a Symbiotic Future*. Palgrave PIVOT series.
2 Plastics Europe. (2022). *Plastics – The Facts 2022*. Plastics Europe. [Online.] https://plasticseurope.org/knowledge-hub/plastics-the-facts-2022/, accessed 12 April 2023.

Problematic Chemistry and Sustainable Development

3

As alluded in the previous chapter, rooted plants have evolved expert chemical means for attraction, defence and communication – bolstered by their close relationships with soil microbes – in part compensating for their inability to move physically and hence to find other means to address their needs. In fact, viewed dispassionately, a serene rural vista is in reality a theatre of intense chemical warfare! In nature, as well as in the human world, chemicals can be beneficial or harmful depending on their recipents and their usage.

3.1 NATURE'S CHEMICAL ARMOURY

The arum lily (*Arum maculatum*) is a widespread, low-growing herb across Europe. It is also a plant associated with many myths and cultural associations, lending it a rich lexicon of common regional names. In England alone, these common names include cuckoo-pint, snakeshead, adder's root, arum, wild arum, lords-and-ladies, devils and angels, cows and bulls, Adam and Eve, bobbins, naked girls, naked boys, starch-root, wake robin, friar's cowl, sonsie-give-us-your-hand, jack-in-the-pulpit, and cheese and toast. Arum lilies produce toxic chemicals in their leaves and flowers to ward off unwanted grazing and other damaging attentions from animals. Aside from its production of toxic substances as a primary defence mechanism, the arum lily also deploys a wider range of chemical tricks to enable it to prosper. Take, for example, the spadix, a large and dark spike growing within the hood-like flower. The spadix not only increases in temperature by as much as 20°C but also attracts

DOI: 10.1201/9781003453949-3

flies, drawn in by chemical trickery as it emits a powerful faecal odour similar to that of animal dung. The flies drawn to this aroma fall into the base of the flower where they are trapped by guard hairs, transferring any pollen adhering to them onto the flower's female parts. The trapped flies are then showered in pollen grains that attach to their bodies. Overnight, the flower's guard hairs relax, releasing the flies to emerge in search of food the next day and thereby to transfer this new load of pollen to other arum lilies. Further chemical ingenuity is deployed as the fertilised female flowers develop into a cluster of berries that mature from green into a bright red colour. These attractive berries are acrid and extremely poisonous to many animals including humans – arum lilies are one of the most common causes of accidental plant poisoning – yet are palatable and harmless to birds, which proceed to disperse the seeds in their faeces after eating the berries. The plant's common name 'lords-and-ladies' derives from the erect spadix in the flower representing the 'lords' and the formation of a cluster of bright berries representing the 'ladies'. These are some of the chemical tricks evolved by a single common temperate species, the arum lily, which is just one example amongst many of the varied chemical ingenuity found in the plant kingdom (Figure 3.1).

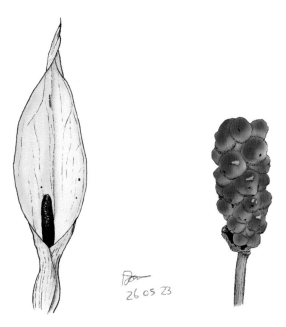

FIGURE 3.1   Arum lily (*Arum maculatum*) flower with spadix and berries, exhibiting examples of nature's evolved chemical ingenuity (©Daisy Everard).

Some chemical uses by plants are to attract other species. We sense this in the rich spring and summer scents emanating from the flowers of wild roses, honeysuckle and many more wild and domesticated plants. Some release their chemical siren calls by day, and others by night depending on the pollinating species they seek to attract. Attractive though many of these scents are to the human senses, others are distasteful such as the odour of rotting meat emitted by *Rafflesia arnoldii*, a rare plant bearing the world's largest bloom that attracts carrion-eaters in the rainforests of Indonesia to convey its pollen. Other chemicals may also offer attractive flavours, though not all attractant chemicals emitted by plants are detectible by human senses. But across this breadth of chemistry, a huge diversity of butterflies, moths, sawflies, bees, bats and other small mammals, wasps, hoverflies and many more creatures besides are drawn in to spread pollen.

Many species of animal too make use of attractive chemical signals. These range from various mammals that mark territories with spraints (faeces) or urine to either attract mates or to deter competitors, through to caterpillars and sea slugs that accumulate in their bodies some of the distasteful or toxic substances from the plants they eat for chemical deterrence. Snakes, spiders, scorpions, sea anemones, jellyfish and many more animals produce venomous chemicals for defence or to immobilise prey, whilst various amphibians, fish and insects are examples of animals producing poisonous substances in or on the surface of their bodies to deter or kill would-be predators. Female moths release pheromone substances into the air to attract potential mates, particularly striking examples found in moth species with totally flightless females that are effectively chemical sirens, unable to move far but luring in flighted and hence mobile male moths from afar.

Although many plants use chemistry to attract specific animals, some produce chemicals with the opposite intent of dissuading interaction. Various plant species generate chemicals that are distasteful or that gum up the mouthparts of would-be grazers. Coffee plants resort to drugging would-be grazing insects by production of caffeine, a chemical that makes these insects shudder so that they fall off the bush; it is exactly this stimulant effect that we humans happen to like when taking a shot of coffee! As we know from the discovery of antibiotics, fungi and other microorganisms are also expert in chemical warfare.

A great many amongst the diversity of static and slow-moving species, including not only fungi and plants but also sessile and slow-moving animals, have evolved ingenious chemical means to kill, dissuade, attract, stimulate, repel or otherwise interact with the myriad life forms with which they have co-evolved. Why would we be surprised that rooted and other immobile organisms, unable either to evade others or to move towards potential mates, have evolved sophisticated alternative chemical means to address their survival needs? The reality is that plants, fungi and other sessile organisms are far more sentient and interactive than commonly assumed, reliant on a bewilderingly broad range of chemical and other stimuli to interact in multiple ways with other organisms of their own and other species.

3.2 HUMAN USES OF CHEMICALS

Evolution of a bewildering array of chemical substances has enabled different wild species to achieve competitive success. The same could be said of human exploitation and innovation of chemistry.

From prehistory, humans have made use of the herbal, poisonous, attractive and other properties of chemicals in the organisms that surround them. These have been deployed as poisons, medicines, stimulants, dyes, perfumes, flavourings, preservatives and many more purposes besides. We continue to make such uses in our increasingly technology-driven world, including, as just one example, taxol – a chemical derived from yew trees and better known by its pharmacological name of Tamoxifen – serving as a powerful, widely used anti-cancer drug that has enhanced the lives of many people around the world, significantly including survivors of breast cancer.

A variety of the world's cultures have embraced this deep relationship with the wider ecosystems of which we are part within their belief systems and practices. This includes, for example, in Ayurvedic medicine practised widely across India, encapsulating centuries of localised knowledge about the properties of plants and their influences on human wellbeing. The rest of us use derivatives of natural medicines in our daily use of drugs, and enjoy the relaxing, stimulating and flavoursome chemicals generated plants as we enjoy a cup of tea or coffee or the fermentation products of microbes in beer or wine, in cheese or yoghurt, and other food and drink. This exotic chemistry is also the basis of myth, legend and art, such as the witches brew in Shakespeare's Macbeth.

We have, of course, made use of nature's chemistry for many more purposes – the preceding section of this book 'A history of humanity through chemical and material use' provides a rapid overview of the breadth of uses – to meet our needs for food production, heat and other requirements for energy, shelter, food, explosives, artistic expression, defence and offence, and many more besides. And let us not forget that synthetic materials such as plastics and pesticides, whilst not formerly encountered in nature, are nonetheless molecules reconfigured from natural sources.

3.3 HUMAN ABUSES OF CHEMICALS

Whilst humans are far from the only species making use of chemistry to meet their needs, the creativity of people in the use of chemicals, the nature of this chemistry and the scale at which this occurs can take us beyond the safe limits beyond which problems are likely to occur. In biospheric terms, this can potentially turn a beneficial use into an abuse.

'Abuse' may perhaps appear too strong a term in this context, but it is more or less accurate if we accept the definition of the word as being that which causes harm to others. And, with a global population exceeding 8 billion people, more than half living 'middle class' lifestyles in the sense of having surplus income that almost automatically drives higher per capita material usage, abuse to the planetary ecosystem though cumulative human pressures is a fair summation of its consequences. When ecosystems are overwhelmed, in terms of what is taken from and what is released into them, their integrity and functioning degrade. And, with that degradation, the capacities of natural systems to support human health, wealth-creating activities, security and opportunity into the future are serially eroded. Degradation of the world's supportive ecosystems through chemical, physical, economic, political or other means, and competitive over-exploitation denying access to and imposing contamination on sectors of society including future generations, is evidently a form of abuse that lies at the roots of many of today's deepening sustainability concerns.

The evidence of ecosystem degradation at global scale is clear and alarming. A stark indicator of this is that some 96% of global mammalian biomass now comprises humans and livestock with only 4% 'wild' species, and also 70% of global bird biomass is now accounted for by poultry raised for human use.[1] The pace of biospheric degradation is also accelerating, with populations of many wild animal species declining by 50% between 1970 and 2018.[2] The average rate of vertebrate species loss over the last century is one hundred times higher than the background rate, indicative that we are in the midst of a sixth mass extinction.[3] Climate change is not only widespread and rapid, but also intensifying.[4] Global resource extraction has more than tripled since 1970, leading to depletion and virtually inevitable future hardships.[5] Even the rising occurrences of novel zoonotic diseases, for example Covid-19 and also MERS, SARS, HIV, West Nile Virus, Zika and many more, are related to the degradation of ecosystems and their regulatory processes, increasing permeability for animal diseases to jump to human hosts.[6] This precipitous decline in the resources and processes of the planetary ecosystem increasingly threatens humanity's life support system and our capacities to live secure and fulfilled lives.[7,8,9]

The abuses of nature witnessed throughout human history and particularly since the Great Acceleration – the unprecedented surge in human numbers and activities since the mid-twentieth century – have arisen as collateral damage in the headlong yet myopic pursuit of progress, rather than resulting from wilful destruction. The accelerating change consequent from a narrowly technocentric and financial pathway of progress is ultimately self-limiting due to oversight of its tendency to erode the foundational natural and human capital upon which this model of 'progress' depends. The question is not whether the resilient planetary ecosystem will survive – it will certainly do so in whatever damaged or modified form – but whether human security and opportunity will be sustained or, alternatively, will continue to be serially undermined. A pathway of sustainable development, or in other words a form of development that can continue indefinitely, is a human-centred world view, albeit one that rests ultimately on the capacities of the natural world to continue to provide a wealth of beneficial processes and services supporting the health and harmonious operation of human society.

Aside from wider political, economic and other forces pertaining to a potentially sustainable world, our use of planetary and novel chemistry, including interactions with natural cycling processes, has a great bearing on current and future wellbeing. A broad range of environmental, health and ethical concerns associated with chemical use has emerged. These range from resource depletion to pollution consequent from releases of substances exceeding nature's capacities to break them down and safely assimilate them, degradation of the physical structure of nature hampering efficient cycling of substances, and inequities in society resulting in unequal access to resources or exposure to risks.

The patterns of use of materials of all types, both natural and synthetic, are as much a problem as the nature of the exploited materials themselves. Poorly controlled mining, felling of forests and input-intensive farming all exert destructive pressures at raw material extraction phases. Manufacturing processes are inherently polluting, though more stringent controls can rein in emissions that may be damaging to human and ecosystem health. Degradation of materials during use can release their constituents, and maintenance inputs during the use phase of products can sometimes substantially outweigh the environmental burdens of material manufacture. A pattern of linear resource use inevitably results in the accumulation of waste when products reach the end of their useful lives, with incautious disposal beyond end-of-life generating pollution whilst wasting valuable resources. These stages in the linear life of products and materials, which have dominated societal habits since the onset of the Industrial Revolution, create sustainability challenges regardless of material type. All types of materials have additional specific challenges of their own, be they ethics in raw material supply chains, pollutants emitted by poor manufacturing practices, inherent non-recyclability, or short useful service life. All have inherent energy costs in manufacture, not forgetting that recycling can also be energy-intensive and so needs to be considered in assessing the sustainability of life cycles. All these factors have to be weighed together in an informed consideration of what is sustainable, what is not, and what the priorities are in moving from today's unsustainable norms towards a goal of sustainability.

The growing consumption of plastics across societal sectors and global regions is problematic in the absence of efficient recovery and recycling of these durable materials. As stated in a 2022 study by SYSTEMIQ, a body founded in 2016 to drive the achievement of the Paris Agreement on climate change and the UN Sustainable Development Goals by transforming markets and business models, "Plastic is both an icon of prosperity and a cautionary example of how linear models of consumption can undermine Earth's planetary limits".[10] A 2017 study estimated that 8,300 million tonnes (Mt) of virgin plastics had been produced to date, generating 6,300 Mt of plastic waste – 9% estimated as recycled, 12% incinerated and 79% accumulated in landfills or the natural environment – and projecting that approximately 12,000 Mt of plastic waste would be in landfills or the natural environment by 2050 without changes in life cycle management.[11] Of this 9% plastic recycling rate, Europe leads the way at 30%, though leaving substantial room for improvement, with 9% recycling in the US and close to zero across much of the developing world. In the absence of improvements in waste management practices, fluxes of plastics entering the oceans was estimated in 2015 as likely to increase by an order of magnitude within the next decade.[12] This accumulation of waste is common to many resistant materials, for example with an estimate that over 3,000 tonnes of aluminium was likely to end up in landfill over Christmas 2019 in the UK alone, largely related to food wrappers that are not recycled.[13] Accumulation of plastics of varying kinds, as well as metals and other potentially useful but also potentially polluting materials entering landfill or the open environment, is symptomatic of the wasteful and careless behaviours of society. It is also indicative of the need to make greater progress with resource recovery for beneficial reuse, reversing global society's profligate and anachronistic linear resource use patterns. The OECD (Organisation for Economic Co-operation and Development) has recognised that "Reducing pollution from plastics will require action, and international co-operation, to reduce plastic production, including through innovation, better product design and developing environmentally friendly alternatives, as well as efforts to improve waste management and increase recycling".[14]

Rising human demand and declining natural capacity creates increasing pressures impinging on the freedom of operation of businesses and society. These mounting pressures will progressively enforce more sustainable behaviours. This has often occurred through expensive retrospective reaction, such as responses to emerging resource depletion or public and regulatory demand changing acceptable norms. However, proactive anticipation and strategic choices to behave in more sustainable ways are less disruptive, as well as potentially profitable, reflecting that unavoidable sustainability pressures will increasingly shape future markets and acceptable behaviours, products and processes. In this regard, with rising human population and demands depending on dwindling natural resources, more rather than less novel chemistry is required better to meet human needs in the most efficient and safest manner. As one example, durable materials can imbue products with a longer service life, potentially also entailing lower maintenance inputs during use, with low wastage and, ideally, with inherent recyclability recapturing material value into products meeting further needs.

A facet of human exploitation of materials that has often been overlooked is that most, be they plastics or non-plastics, require additives at different stages of their life cycles to provide or maintain function. Additives to all types of plastics are well documented. But timber treatments are commonplace, be that at the point they are converted into useful products, such as pressure treatment with biocidal preservatives, or else periodic additions of preservatives during product use in order to extend useful service life. Coatings of various types are commonly applied to metal, glass, card and paper to enhance aspects of their technical performance. Flame retardants and treatments with pest repellents or biocides are necessary additives for various organic and other long-life insulation materials. Glazes are added to pottery. Energy, chemical and labour inputs are required to maintain the serviceability of many products throughout their lives. These additives to product manufacture or maintenance are often overlooked in more simplistic assessments of environmental or sustainability footprints, yet can be substantial when the whole life cycle of material use is considered.

International sustainable development charity The Natural Step (TNS) uses the metaphor of 'The Funnel' to describe the situation of declining environmental capacity impinging on society's freedoms to operate. As historic examples, slavery was once an efficient means to reduce labour costs promoting profitable business, but fell foul of shifting public values and acceptance resulting in a ban as it 'hit the walls of the funnel'. The use of lead in paint or as a vehicle fuel additive may have yielded technical efficiency, but the consequences of lead poisoning particularly on infant development led progressively to bans as another example of environmental, health and political factors combining to 'hit the walls of the funnel', requiring innovation of less toxic novel approaches more aligned to that aspect of sustainability. A warning note has, however, to be sounded. If businesses, regulators and governments rely on reactivity only when inherently problematic issues become evident, acting only after the weight of publicly available evidence becomes sufficiently undeniable to exceed the often-substantial inertia created by powerful vested interests, retrospective action may be too late and hugely expensive. It may also result in 'knee-jerk' reactions to phase out a substance or process of current concern, substituting it with another that has not been rigorously assessed and that may therefore merely lead to the locking of investment into 'tomorrow's problem'. Foresight informed by knowledge of sustainable development and the pressures it may pose in future – a level playing field of principles applied to all material choice and innovation choices – is an essential foundation for sound decision-making.

We see abundant evidence of the lag effect of vested interests today in the rhetoric around climate change wherein, in the face of substantial evidence about the need urgently to constrain climate-active gaseous emissions, proportionate action to decarbonise the economy is held back at great future risk by those benefitting from maintaining the status quo. This is particularly the case by actors in the petrochemical, vehicle and livestock sectors, and by producing countries, cumulatively with sufficient weight of economic power and political influence to lobby against, delay or overturn decision-making.[15] Furthermore, reliance on reaction only once problems arise, akin to often ineffective attempts to quell the spread of invasive alien species of plants and animals once established, may result in too little action, too late, when damage is not only already substantial but is potentially irreversible.

3.4 SOCIETAL RESPONSES TO CHEMICALS OF POTENTIAL CONCERN

Manufacture and use of a diverse array of chemicals have preceded awareness of potential issues of concern. Sometimes, wider public disclosure of problems that were known within businesses has been suppressed in the interests of profit-making, clear examples being risks associated with the use of tobacco, prevalence and impacts of some 'forever chemicals', widespread use of asbestos and seriously adverse impacts of neonicotinoid pesticides on wildlife.[16,17]

The 'environment movement', as we regard it today, began to take tangible and increasingly vocal form in the early 1970s in response to a growing body of evidence of adverse environmental and public health issues in preceding decades. The preceding technocentric culture placed faith in technology for providing social progress, though in the absence of systems thinking and a precautionary approach.[18] This comfortable, economically led world view was to be severely disrupted by evidence of serious issues relating to incautious chemical innovation and use. Perhaps the best known amongst many revelations communicating public and political risks associated with formerly unforeseen adverse environmental and human health consequences of chemical use was Rachel Carson's 1962 book *Silent Spring*.[19] This seminal book collated knowledge about the global prevalence and implications of bioaccumulating pesticide substances, formerly seen as a public good by eradicating problematic pest organisms and increasing food security but with far wider ramifications for biodiversity and human health. The revelations of *Silent Spring* were to drive presidential initiatives in the US leading to the banning of DDT within a decade, and precipitating wider global concern about the uncontrolled use of persistent substances in the open environment. 1972 was a milestone in recognition of human interdependency with the global ecosystems, with the UN 'Stockholm Conference' (the UN Conference on the Human Environment[20]), instigation of the United Nations Environment Programme (UNEP) and publication of the Club of Rome's 'Limits to Growth' report.[21] In the US, the Environmental Protection Agency (EPA) was established in 1970 under the administration of President Richard Nixon to create consistent national guidelines and enforcement, replacing a wide range of confusing and often ineffective environmental protection laws enacted by states or by communities. The EPA was initially charged with administering the US Clean Air Act (1970), but its remit progressively expanded in scope to address the authorisation and regulation of pesticides, responsibility for clean water including the regulation of municipal and industrial effluents, control and clean-up of waste sites and, in the early twenty-first century, addressing climate change. However, retrospective controls on substances in use before 1974 has been slow to be enacted, meaning that many such substances have been authorised under a 'grandparenting' approach that lacks systemic scrutiny.

The regulation of chemicals across the world is uneven with respect to how they deal with risks. As one example, the EU REACH (Registration, Evaluation, Authorisation and Restriction of Chemicals) regulations, although claimed to take a risk-based approach, are substantially informed by the 'intrinsic properties' of substances under consideration. 'Intrinsic properties' relate to hazard alone, with risk considerations under REACH only applied once a substance has been scheduled as a candidate for substitution. Hazard assessment alone can be misleading as it lacks context, and so consequently overlooks risk. Water, for example, is inherently not only safe but essential for life, yet is the medium in which an estimated 236,000 annual drowning deaths occur worldwide, constituting the third leading cause of unintentional injury deaths.[22] As scuba divers know, oxygen can turn into an instantly fatal neurotoxin beyond certain partial pressures, and the inert gas nitrogen also turns into a narcotic at depth. Context matters not only in terms of the use of 'safe substances', but also in terms of those with more hazardous inherent properties yet that may be wholly consumed in enclosed production processes, or that may be immobilised and inert in products that are then recycled in closed technological cycles at end-of-life. Regrettably, the regulatory environment has not yet caught up with assessment of context-specific risk.

A further consequence of this regulatory naivety is reinforcement of over-simplistic assumptions that there are inherently 'good' and 'bad' materials. Life cycle context is key. Take for example a biologically sourced timber product, often lauded as inherently sustainable if sourced from well-managed forests. Yet the biodegradable nature of timber may mean that it has to be regularly treated with protective and often biocidal substances to prolong the useful service life of timber products, after which the wood cannot be safely composted to return it to nature; in fact, it generally then constitutes contaminated waste. Persistent substances may have a bad reputation under a simplistic good/bad framing, and this is reasonable in the context of a persistent pesticide applied in the open environment of a field where it may disperse and also accumulate in soils, water, nature and people remote from source. However, in a durable product, such as a window profile or pipe, a 'persistent' plastic may be chemically inert whilst providing very long service life with little or no inputs of materials or energy for maintenance, providing a high degree of insulation and other services, and after which it may be recycled to capture its material value and embodied energy whilst also avoiding waste and providing further beneficial use. In these contrasting examples, durability, a facet of persistence, can have strongly positive pro-sustainability virtues in delivering a high level of societal benefits with minimal environmental footprint in maintenance and remanufacture. The inherent recyclability as well as the existence of operational recycling infrastructure are highly germane to overall life cycle sustainability footprint, as are the environmental and social footprints associated with the raw material supply chains serving the manufacture of these substances.

There is, in reality, no such thing as an inherently sustainable material. However, taking account of wider life cycle contexts, there is such a thing as sustainable USE of materials, a topic to which we will return later in this book. There is a pressing need for a more systemic and common 'level playing field' approach to assessment of the sustainability of the use of chemicals, informed by the context of the 'real world' societal life cycles of the products into which they are incorporated.

A particularly effective strategic approach to sustainability assessment that has been applied to the use of chemicals is that advocated by international sustainable development charity TNS. The TNS approach is based on understanding the scientific principles of what constitutes a sustainable future, informing stepwise progress through the process of backcasting from an idealised vision of a business, value chain or other enterprise or product fully compliant with those principles. Reaching for perfection in terms of that sustainable vision may not be feasible immediately or even, from the starting point of a largely unsustainable present, in the short or medium term. However, incremental decision-making and investments that constitute currently feasible 'stepping stones' leading towards a clearly articulated sustainable goal can result in wise strategy that averts future, formerly unanticipated 'shocks' that can hamper profitability and public trust. The alternative is making 'knee-jerk' decisions based on media or NGO pressure that can result in sinking investments in alternative materials or processes that are not tested in terms of their fitness for sustainability, likely to lock the enterprise into poor and risky decisions over long return-on-investment periods. As an example of a non-strategic 'pro-sustainability' decision, the 1987 intergovernmental Montreal Protocol to phase out the CFC (chlorofluorocarbon) substances implicated in damage to the planetary ozone layer was a necessarily good thing. However, corporate pressure to replace CFCs with their available HCFC (hydrocholorofluorocarbon) alternatives was not a wise move when viewed from a backcasting perspective, as the ozone-depleting potential of HCFCs was still significant, if lower than CFCs, but the climate-forcing impacts of HCFCs were greater. It is therefore essential, in thinking about the sustainable use of chemicals, as well as other substances or processes, to base thinking in a science-based and systemic framework in which the principles of sustainability are clearly articulated.

This was the purpose behind the instigators of TNS in Sweden in the late 1980s. Today, TNS operates internationally, and many of its operating principles can be seen more widely implemented in other approaches to sustainable development. The late 1980s was a period of great concern about environmental factors undermining ecosystem security and public health. It was also a period in which complex arguments raged about the virtues of the use of competing families of chemicals, the relative importance of different organisms, the rights of different societal groups, and other facets that tended to lead to dissonance rather than consensus. The building of consensus is deeply rooted in Swedish culture, perhaps more so than in any other country at least in Europe, and it was in recognition of a need for an unambiguous framework of thinking about sustainable development that was portable across societal sectors (including business) and across scales that the Swedish oncologist Dr Karl-Henrik Robèrt embarked on a consensus-building process across a broad spectrum of scientific disciplines and societal sectors in Sweden. Cutting through the layers of contested science, Robèrt based a model of the workings of the planetary biosphere, and of cellular metabolism within it, upon non-contentious, basic scientific principles – the laws of thermodynamics, the principle of matter conservation and so on – upon which all could agree.

The disarmingly simple TNS model of biospheric cycles emerging from this process is not dissimilar to one that most of us will have learned at school. Matter is maintained in circulation by biological processes driven by the capture of solar energy by plants that use it to build complex organic substances, which are then broken down by an array of organisms in complex ecosystems liberating energy and excreting simple 'building block' molecules that are again reintegrated by plants: a cyclic system with no waste matter. However, reflecting the process of planetary evolution, a second cycle operating over geological timescales is recognised in the TNS model. The Earth protoplanet initially constituted a largely homogeneous cloud of gases that subsequently condensed into discrete layers of rock, water and atmosphere. Physicochemical processes, such as deposition and sequestration in deep oceans, were subsequently accelerated by biomineralisation processes following the emergence of life, resulting in the 'locking away' of many substances formerly in free circulation into the developing crust of the planet. Many of these sequestered substances – including for example heavy metals and high concentrations of carbon and nutrient substances – are toxic to current life forms or disruptive to contemporary planetary cycles when present at their previously elevated concentrations.

The next step in development of the TNS systems science model was to recognise four key, and again non-contentious, principles upon which the system depends. These four principles were reframed in the negative: if we do not want to break the sustainable Earth system then we must not do these things. These were worded as the TNS 'System Conditions', defining what we must not do if we are to maintain or achieve a sustainable Earth system. Figure 3.2 is a simple representation of the TNS Systems science model, and Table 3.1 outlines the four TNS System Conditions with examples of the kinds of issues to which each relates.

The TNS systems science model, the four TNS System Conditions, 'The Funnel' metaphor and a four-stage A, B, C, D approach to backcasting are

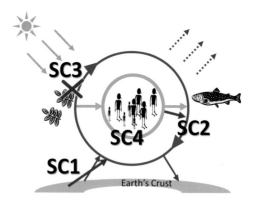

FIGURE 3.2   The TNS systems science model, indicating the four TNS System Conditions (SCs) (©Mark Everard).

TABLE 3.1   The four TNS System Conditions with examples of relevant issues

TNS SYSTEM CONDITION	*EXAMPLES OF RELEVANT ISSUES*
IN THE SUSTAINABLE SOCIETY, NATURE IS NOT SUBJECT TO SYSTEMATICALLY INCREASING…	
1.... concentrations of substances extracted from the Earth's crust	Release of mined substances at rates exceeding nature's re-assimilation rates can result in systematic accumulation with unpredictable consequences, as for example releases of mined carbon-rich deposits, heavy metals, radioactive substances and phosphorus
2.... concentrations of substances produced by society	Release of substances alien to nature produced by society may not be broken down or re-assimilated at rates preventing their tendency to systemically accumulate with unpredictable consequences, including for example persistent pesticides and other substances, CFCs/HCFCs and various 'forever chemicals'
3.... degradation by physical means	Systematic overharvesting or conversion of nature's productive infrastructure can perturb natural cyclic processes, such as high rates of forest felling, fishery, 'green cover' and soil depletion, and overharvesting from aquifers and surface waters beyond natural replenishment rates
AND, IN THAT SOCIETY	
4.... there are no structural obstacles to people's health, influence, competence, impartiality and meaning	The organisation of society can accelerate pressures on resource depletion or systematic accumulation, undermining security and opportunity in the present and/or future for different sectors of society such that people are prevented from meeting their needs

part of a wider suite of TNS tools collectively known as the Framework for Strategic Sustainable Development (FSSD). The FSSD collectively constitutes a methodology supporting decision-makers, institutions and other societal sectors to develop a vision of strategic goals informed by science-based sustainability principles. Backcasting from this visionary end-point enables the identification of currently feasible yet strategic steps that lead in the direction of the sustainable end-goal, and that may also serve as enabling platforms for further steps, and the rational evaluation of trade-offs. (These steps may, in the context of the societal use of chemicals, include chemical innovations, improved management such as containment of potentially hazardous substances in manufacturing, development of sorting and recycling infrastructure and user behaviours to support more circular use, enabling revisions in the policy and/or fiscal environment, or other means to lead to the eventual goal.)

Comprehensive and regularly updated details about TNS can be found on the TNS organisation's website (https://thenaturalstep.org). Many examples of its use internationally by businesses, municipalities and other societal sectors can also be found on that website, including in the construction sector, aviation, hospitality, food systems, material science, domestic appliance, apparel and many more applications. This overview of the TNS approach is important not only to illustrate a systems context for getting to grips with 'real world' sustainable development challenges, but also as TNS principles have played a key role in the engagement of the European PVC sector.

3.5 SERVING HUMAN NEEDS

In closing this consideration of problematic chemistry, sustainable development and some approaches to sustainability assessment, it is important to consider the reason that we make and use chemicals in society.

Based on the focus of the bulk of chemical regulatory approaches and tools around the potential of chemicals to be harmful, a casual observer might surmise that businesses innovate chemicals to cause harm. The metrics used in European approaches such as EU REACH, Product Environmental Footprint (PEF) and Environmental Product Declaration (EPD), as well as international approaches such as Life Cycle Assessment (LCA), essentially catalogue such harmful factors as ozone depletion, ecotoxicity, CMR (carcinogenic, mutagenic and reprotoxic properties) and eutrophication potential. Whilst these are important factors to consider in sustainability assessment, they are, of course, not why we use chemicals, nor are they why organisations innovate and market them. Organisms use a variety of chemicals to serve their needs, and people are no different in exploitation of natural or novel chemicals to serve their diverse survival, economic and other needs as we observed when considering the history of humanity through chemical and material use. Whilst not discounting the vital importance of accounting for associated risks, a key focus of chemical use and innovation should also be on how they enable us to achieve benefits and meet needs.

The achievement of near-global consensus in 2015 about the UN Sustainable Development Goals (SDGs), including their associated 169 targets as guidance for sustainable development to 2030, was hugely helpful by refocusing sustainable development efforts on addressing human and environmental needs. This reframing on benefits rather than, as generally transposed in regulations and public psyche, the negative elements of substances and processes, has been recognised as reasserting the 'lost half' of sustainable development in the form of the primacy of human needs as end-points.[23] Chemical usage and further innovation, amongst a range of other societal activities and interests, should be evaluated in balanced terms of what best meets human and environmental needs in the safest and most environmentally efficient manner. This reorientation of the concept of sustainable development, relative to what had become accepted as a norm, is in fact a realignment with the boldly intergenerational framing of the 'Brundtland definition',[24] which addressed meeting human needs now and into the future. A reframing on the meeting of needs opens up dialogue about opportunity and innovation, stimulating business creativity and recognising novel future markets under a bold re-articulation of the necessary paradigmatic change.

In the world of chemical usage, this reframing of purpose towards optimal means to meet needs can better recognise the positive contribution of material durability, for example in long-lived infrastructure that may also be recovered for recycling and further beneficial reuse, maximising human utility whilst minimising material content and maintenance inputs. A positive framing on needs has, of course, still to be balanced against unwise uses of persistent substances likely to lead to systematic accumulation in ecosystems and human cells as, for example, in the case of uncontrolled use of dangerous pesticides or the accumulation in ecosystems of single-use plastic bags and other articles. Chemicals conferring low maintenance also make substantial savings on associated inputs of staff time, energy and other substances that may sometimes exert a greater sustainability footprint than the production of the materials themselves. Recyclability is a further asset, recovering material and energy value with greatly reduced reliance on virgin inputs and waste generation with their associated environmental and social consequences. Optimally, novel materials meeting human needs should also have flexible characteristics that better adapt them to meet those needs and to enhance durable and cyclic use, whilst also being accessible for more people to meet their various needs through greater affordability.

There are parallels in this paradigm shift in chemical assessment with progressive thinking about progress with managing emissions of carbon and other climate-active gases. Transition from a narrow focus on potential chemical hazard, through to consideration of benefits in terms of meeting needs, has interesting commonalities in changing thinking about climate-active emissions from a 'carbon footprint' to a 'carbon handprint' approach.

The carbon footprint approach essentially audits the total greenhouse gas emissions of an individual, or an event, organisation, process, product or service, generally expressing this as a carbon dioxide equivalent (CO_2e) mass. In other words, a carbon footprint audits how bad the associated emissions are. Carbon footprint has important uses, but can also be a cause of guilt and low motivation. Over recent years, there has been growing interest in an alternative or additional carbon handprint approach, focusing on the positive contributions that people and institutions can make towards reducing carbon emissions recognising that no contribution is too small. The growing commitment and appetite of governments at all scales, businesses, civil institutions and citizens to do more to address looming climate threats are served in a far more motivating way by recognising all of these positive contributions, emphasising that anybody can make a difference through their efforts. By recognising and auditing positive contributions, the carbon handprint approach is a motivator of individual and collective action towards reducing climate-active emissions as a positive impact on the environment. The innovation of products, processes and services that help others reduce their carbon footprint also count as part of the carbon handprint of an organisations or an individual. When carbon handprint and carbon footprint are equal, carbon neutrality is attained; when handprint exceeds footprint, climate positivity is achieved. The carbon handprint approach positively acknowledges all progress on the journey towards those generally more aspirational goals.

NOTES

1 Bar-On, Y.M. and Phillips, R. (2018). The biomass distribution on Earth. *PNAS*, 115(25), pp. 6506–6511. DOI: https://doi.org/10.1073/pnas.1711842115.
2 WWF. (2018). *Living Planet Report 2018: Aiming Higher.* Worldwide Fund for Nature (WWF). [Online.] https://www.panda.org/knowledge_hub/all_publications/living_planet_report_2018/, accessed 25 January 2022.
3 Ceballos, G., Ehrlich, P.R., Barnosky, A.D., García, A., Pringle, R.M. and Palmer, T.M. (2015). Accelerated modern human–induced species losses: Entering the sixth mass extinction. *Science Advances*, 1(5), p. e1400253. DOI: https://doi.org/10.1126/sciadv.1400253.
4 IPCC. (2021). *Climate Change Widespread, Rapid, and Intensifying.* Intergovernmental Panel on Climate Change (IPCC), 09 August 2021. [Online.] https://www.ipcc.ch/2021/08/09/ar6-wg1-20210809-pr/, accessed 18 March 2023.
5 UNEP. (2019). *We're Gobbling Up the Earth's Resources at an Unsustainable Rate.* United Nations Environment Programme (UNEP). [Online.] https://www.unep.org/news-and-stories/story/were-gobbling-earths-resources-unsustainable-rate, accessed 18 March 2023.

6 Everard, M., Johnston, P., Santillo, D. and Staddon, C. (2020). The role of ecosystems in mitigation and management of Covid-19 and other zoonoses. *Environmental Science and Policy*, 111, pp. 7–17. DOI: https://doi.org/10.1016/j.envsci.2020.05.017.

7 Díaz, S., Fargione, J., Chapin, F.S. and Tilman, D. (2006). Biodiversity loss threatens human well-being. *PLoS Biology*, 4(8), pp. 1300–1305. DOI: https://doi.org/10.1371/journal.pbio.0040277.

8 Cardinale, B.J., Duffy, J.E., Gonzalez, A., Hooper, D.U., Perrings, C., Venail, P., Narwani, A., Mace, G.M., Tilman, D., Wardle, D., Kinzig, A.P., Daily, G.C., Loreau, M. and Grace, J.B. (2012). Biodiversity loss and its impact on humanity. *Nature*, 486(7401), pp. 59–67. DOI: https://doi.org/10.1038/nature11148.

9 Hooper, D.U., Adair, E.C., Cardinale, B.J., Byrnes, J.E.K., Hungate, B. A., Matulich, K.L., Gonzalez, A., Duffy, J.E., Gamfeldt, L. and O'Connor, M.I. (2012). A global synthesis reveals biodiversity loss as a major driver of ecosystem change. *Nature*, 486(7401), pp. 105–108. DOI: https://doi.org/10.1038/nature11118.

10 SYSTEMIQ. (2022). *ReShaping Plastics: Pathways to a Circular, Climate Neutral Plastics System in Europe*. SYSTEMIQ. [Online.] https://www.systemiq.earth/wp-content/uploads/2022/04/SYSTEMIQ-ReShapingPlastics-ExecutiveSummary-April2022.pdf, accessed 12 April 2023.

11 Geyer, R., Jambeck, J.R. and Law, K.L. (2017). Production, use, and fate of all plastics ever made. *Science Advances*, 3(7), p. e1700782. DOI: https://doi.org/10.1126/sciadv.1700782.

12 Jambeck, J.R., Geyer, R., Wilcox, C., Siegler, T.R., Perryman, M., Andrady, A., Narayan, R. and Law, K.L. (2015). Dumping lots of plastics into our oceans. *Science*, 347(6223), pp. 768–771. DOI: https://doi.org/10.1126/science.126035.

13 Circularonline.org.uk. (2019). *Over 3,000 Tonnes of Aluminium Will End Up in Landfill this Xmas*. Circularonline.org.uk, 3 December 2019. [Online.] https://www.circularonline.co.uk/news/over-3000-tonnes-of-aluminium-will-end-up-in-landfill-this-xmas/, accessed 12 April 2023.

14 OECD. (2022). *Plastic Pollution is Growing Relentlessly as Waste Management and Recycling Fall Short, Says OECD*. Organisation for Economic Co-operation and Development (OECD). [Online.] https://www.oecd.org/environment/plastic-pollution-is-growing-relentlessly-as-waste-management-and-recycling-fall-short.htm, accessed 12 April 2023.

15 Carter, L. and Dowler, C. (2021). *Leaked Documents Reveal the Fossil Fuel and Meat Producing Countries Lobbying Against Climate Action*. Unearthed, Greenpeace, 21 October 2021. [Online.] https://unearthed.greenpeace.org/2021/10/21/leaked-climate-lobbying-ipcc-glasgow/, accessed 13 April 2023.

16 EEA. (2001). *Late Lessons from Early Warnings: The Precautionary Principle 1896–2000*. Environmental issue Report No 22. European Environment Agency (EEA), Copenhagen. [Online.] https://www.eea.europa.eu/publications/environmental_issue_report_2001_22, accessed 18 March 2023.

17 EEA. (2013). *Late Lessons from Early Warnings: Science, Precaution, Innovation:* Summary. EEA Report No 1/2013. European Environment Agency (EEA), Copenhagen. [Online.] https://www.eea.europa.eu/publications/late-lessons-2, accessed 18 March 2023.

18 Rutherford, M. (2004). *Institutions in Economics: The Old and the New Institutionalism.* Historical Perspectives on Modern Economics. Cambridge University Press.
19 Carson, R. (1962). *Silent Spring.* Houghton Mifflin Company. Cambridge (Massachusetts): The Riverside Press, 1962.
20 United Nations Conferences. (1972). *UN Conference on the Human Environment, 5–16 June 1972, Stockholm.* United Nations Conferences. [Online.] https://www.un.org/en/conferences/environment/stockholm1972, accessed 14 March 2023.
21 Meadows, D.H., Meadows, D., Randers, J. and Behrens, W.W. III. (1972). *The Limits to Growth.* Potomac Associates – Universe Books.
22 World Health Organization. (2021). *Drowning.* World Health Organization, 27 April 2021. [Online.] https://www.who.int/news-room/fact-sheets/detail/drowning, accessed 11 March 2023.
23 Everard, M. and Longhurst, J.W.S. (2018). Reasserting the primacy of human needs to reclaim the 'lost half' of sustainable development. *Science of the Total Environment*, 621, pp. 1243–1254. DOI: https://doi.org/10.1016/j.scitotenv.2017.10.104.
24 World Commission on Environment and Development. (1987). *Our Common Future.* Oxford University Press.

PVC
The Good, the Bad and the Prejudiced

4

Humans are inveterate users and innovators of materials of all types, from the natural to the synthetic. This book is in essence about chemical and materials usage in its broadest sense. However, within that, there is a focus on polyvinyl chloride (PVC).

Why PVC? Well, PVC is perhaps the most heavily scrutinised bulk chemical substance in use globally today, around which there has emerged a pronounced polarisation of opinion. But, let's recall, it is not a substance distinct from all others; it is just one amongst a bewildering array of chemicals of all types, including a huge diversity of plastic materials, put to use by people to meet various of their needs.

4.1 WHAT IS PVC?

In short, PVC is a polymer known as polyvinyl chloride. The polymer appears to have been invented on three independent occasions. The first was in 1838 by the French physicist and chemist Henri Victor Regnault, who observe a white solid formed in a vessel from vinyl chloride monomer gas. However, it was not until 1913 that Friedrich Klatte rediscovered the polymer, and was granted a patent for PVC that also covered a polymerisation process using sunlight.

Little use was made of this third discovery until the patent lapsed in 1926, opening up further investments in commercial innovations and products by other companies. A significant step forwards in the utility of PVC was the addition of additive substances to stabilise this white material. A further

DOI: 10.1201/9781003453949-4

innovation widening the potential commercial use of PVC was the inclusion of additives to plasticise the polymer, creating a more durable, mouldable and elastic compound. This led to the first commercial applications as a rubber replacement, initially in uses such as shoe soles, wire coverings and tool handles. A further virtue of these new PVC compounds was that they were cheap, formed essentially from chlorine derived from sea water and with the organic content from petroleum.

Diversification of PVC compounds mixed with a range of additives have substantially expanded the material's properties and breadth of applications since the 1930s, and with it a broad pervasion of uses across society. PVC compounds are also widely referred to as 'vinyls', a term derived from the Latin word *vinum* meaning 'wine'. (The 'wine' term was initially associated with ethylene, a prime constituent in the manufacture of the vinyl chloride constituent of PVC as well as of ethyl alcohol which is the type of alcohol present in wine.[1])

There are a number of different PVC manufacturing pathways. The most common, the 'ethylene process', includes reacting ethylene with chlorine to form ethylene dichloride (EDC). (Chlorine is an important co-product of caustic soda production, along with hydrogen which serves many beneficial purposes including increasing interest as an alternative fuel.) The EDC is then cracked to generate vinyl chloride monomer (VCM) and hydrogen chloride. VCM is subsequently polymerised, forming the polyvinyl chloride polymer. Another process, known as the 'acetylene carbine process', is still used particularly in China, largely driven by the country's reliance on its rich coal reserves, but has a far more carbon-intense and 'dirtier' footprint that the ethylene process (Figure 4.1).

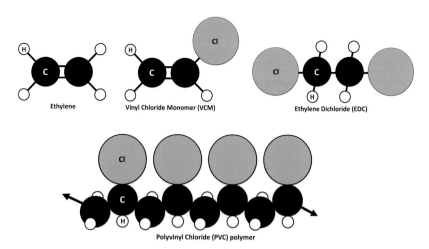

FIGURE 4.1   The molecular structures of ethylene, vinyl chloride monomer (VCM), ethylene dichloride (EDC) and polyvinyl chloride (PVC). ©Mark Everard.

The PVC polymer, as a resin, is compounded with an extremely wide range of additive substances to broaden the properties of the plastic. Three additive categories are essential for maintaining the integrity of the plastic. The three essential functional additive categories are as follows:

- **Stabilisers**. Heat stabilisers, often metallic soaps (metallic salts of fatty acids), are the most significant and earliest additive inventions. Stabilisers prevent the breakdown of polymer chains and links between them, giving the compound a high level of durability as well as many other useful characteristics such as resistance to high heat. Some stabilisers can also confer light stabilisation and ultraviolet stabilisation, which are useful in some PVC applications.
- **Lubricants**. Lubricants can be either external, reducing adhesion between PVC and metal surfaces, or internal, reducing frictional forces between PVC molecular chains thereby reducing melt viscosity. Some lubricants have both properties. Other additives, such as some heat stabilisers and antioxidants, may also have lubricant properties.
- **Plasticisers**. Plasticising substances are widely used in polymer materials, including PVC, to make them flexible. PVC compounds without the addition of plasticisers are known as 'rigid PVC', or PVC-U or uPVC where the 'U' indicates 'unplasticised'. Plasticisers introduced into the polymer compound increase its flexibility.

A wide range of optional additives that are not strictly essential for the integrity of the plastic are used to alter the properties of the compound. These optional additives are added to different polymer formulations to influence the characteristics of the plastic compound, adapting it to specific applications. Some significant optional categories of additive that may be incorporated during compounding include the following:

- **Processing aids**. Processing aids, mainly polymeric acrylates, are added to PVC at low concentrations to enable melt processing of rigid PVC and to expand processing windows. Some processing aids may also have lubricating properties.
- **Antioxidants**. These substances, as the name implies, arrest or slow degradation by inhibiting oxidation. They are used in many types of plastic polymers.
- **Pigments**. Pigment substances add colour and may be used for this purpose in all types of polymer, including PVC compounds as well as many non-plastic materials. Pigments may also play roles in protection against ultraviolet or light breakdown.

- **Impact modifiers**. Impact modifiers are large and generally flexible organic polymers, mainly polymeric acrylates, that are added to PVC and other plastic compounds to confer strength and resistance to cracking.
- **Fillers**. Many PVC applications, such as in flooring, may have a high content of chalk or other inert filler materials to bulk out the plastic compound and to offer it additional performance characteristics. Fillers are common additives in the compounds of other polymers as well as non-plastic materials.
- **Flame retardants**. PVC, unlike many polymers, is naturally self-quenching as chlorine atoms released on combustion suppress fire. Flame retardants are commonly used in other more combustible plastics, such as polyolefins, but are added to some PVC compounds used in specific high-risk applications, especially to suppress smoke emission.
- **Antistatic agents**. As all types of plastic are insulating materials, they are prone to build-up and discharge of electrostatic energy. Antistatic agents control the accumulation of static electrical charge, especially on polymer surfaces.
- **Blowing agents**. These substances may be added to PVC and other types of polymer during compounding to produce a cellular structure when this property is required for certain applications.
- **Compatibilisers**. Compatibilisers are used to promote adhesion between the interfaces of pairs of polymer substances that are otherwise immiscible.

The finished PVC compound can then be converted into a broad range of useful products through manufacturing processes including extrusion, injection moulding, calendering (compressing between heated rollers into continuous film and sheet), blown film production, blow moulding, thermoforming, rotation moulding, and impregnation of fabric to form leather cloth.

This range of properties and processes creates a highly versatile family of plastics that are inert and highly resilient. The fact that these finished plastics are readily cleaned or sterilised ideally suits them to medical applications, as well as water pipes and a range of other uses. PVC compounds are waterproof as well as resistant to heat, acids, salts, bases, fats and alcohols, and can be compounded in rigid or flexible forms. Permeability to gases is low, and this is particularly important in some EU countries, such as The Netherlands, where hydrogen is distributed via PVC piping, and where the leak tightness of PVC fittings has been found to be acceptable for hydrogen distribution.[2]

@ Leak tightness of

PVC fittings with hydrogen.

PVC compounds also have high thermal and electrical insulating properties. PVC is also eco-efficient as only 43% of the molecular weight of PVC polymer is derived from oil with a correspondingly low carbon content, 57% by mass comprising chlorine atoms derived from salt. The potentially long service life of the polymer in useful products and its inherent mechanical recyclability and substantial delivery of utility relative to the mass of molecules, confers additional sustainability benefits.

On the basis of these properties, PVC is found widely in societal applications today. These range from landline and mobile telephones and more widely in computing, information technology and communications infrastructure, in entertainment systems including vinyl records as well as other media and reproduction technology, as insulation, in packaging, widely in aviation and healthcare, in transportation, plumbing, power generation and supply network, in toys, as geomembranes, in renewable energy technologies, in wires and window profiles and in many other construction applications.

After polyolefins (principally polyethylene and polypropylene), PVC is the second-most manufactured polymer globally. Across the world, 38 million tonnes of PVC was produced in 2015, 69% of which is used in the building and construction sector reflecting its suitability for durable applications.[3] The properties of PVC have also resulted in its pervasive use in medical applications such as blood bags, equipment casings, wiring, pipes, implants, cannulas and catheters and wipe-clean surfaces.

This simple overview of PVC suffices for the purpose of this book, the primary focus of which is on sustainable development issues. A great deal more information about PVC can be found from a wide range of online and printed sources.

4.2 THE DARK SIDE OF PVC

So, for such a useful and ubiquitous material, why has PVC been the focus of a substantial amount of NGO and media pressure? Three principal factors have raised concerns: the chlorine content; the diversity of additives; and the resilience of the material after end-of-use. We will look at all three factors in turn.

Perceived ills associated with chlorine manufacturing plants and organochlorine chemicals in general have their roots in a range of uses to which the substance is put. Chlorine gas was used as a chemical weapon during the First World War; its use as a weapon banned under the 1993 Chemical Weapons Convention[4] though its continued use has been reported in the Syrian civil war in 2015.[5] The innovation of synthetic chlorinated organic substances gave rise to a variety of new uses, some families of these substances giving rise to issues of major concern. Amongst these is the use of organochlorine substances as nerve gases. Referring back to Rachel Carson's seminal 1962 book *Silent*

Spring, the ills of bioaccumulation of persistent organochlorine and other pesticides were raised to public and political attention, along with increasing evidence of their contributions to mammalian health issues including cancers, birth defects and direct toxicity, amongst a broader range of environmental impacts. This led to the successive banning or restrictions of open use of a range of organochlorine pesticides. We have encountered CFC (chlorofluoro-carbon or freon) substances already in this book, persistent molecules used as refrigerants and propellants, with the unfortunate unintended consequence of carrying halides into the upper atmosphere where, when split by ionising radiation, they release chlorine free radicals that catalyse the breakdown of the protective ozone layer. The threat posed by CFCs led to their eventual global phase-out, negotiated under the 1987 Montreal Protocol. Other organochlorine substances of concern are PCBs (polychlorinated biphenyls), a class of synthetic chlorinated isomers of biphenyl $(C_6H_5)_2$ valued for their electrical insulation and fire-resistant qualities but discovered to present serious environmental hazards as highly persistent, bioaccumlative and toxic agents. Production and use of PCBs have been limited by international agreement since 1973, but residues of PCB chemicals are likely to remain in the environment for decades or even for centuries.

PVC then is not without the potential for problems. In fact, aspects of the performance of PVC industry were, in the 1980s and into the 1990s, problematic in terms, for example, of releases of vinyl chloride monomer and other organochlorine substances under formerly less well-managed manufacturing processes. Significant manufacturing improvements have been made across Europe and elsewhere as problems became acknowledged, as outlined later in this book. However, this same commitment to improved or best practice is far from global at present. In the use phase of products, the chlorine content of the PVC polymer does not represent a risk as it is bound into the inert molecular structure of the plastic. Potential risks can still arise from the generation of chlorinated substances, such as hydrogen chloride or more persistent compounds (for example dioxins) at end-of-life where incautious disposal such as open burning is permitted. However, an interesting question arises about relative risks of liberation of chlorine-containing compounds from PVC in relation to other substances if burning occurs in less than ideal conditions that fail to break down problematic combustion products. Chlorine is present in sea water, blood and living cells, and in many naturally occurring substances including as an active constituent of chlorophyll in plant matter as well as in most of the food we eat. Consequently, the chlorine content even of apparently inert waste is not limiting to the formation of potentially problematic organochlorine substances if poor incineration practices occur. Disposal methods rather than inherent chemistry lie at the root of this issue of concern. Management of materials when products reach end-of-life is a wider societal problem, and one for which concepts such as the 'circular economy' are being formulated across Europe, with parallels elsewhere in the world.

The use of some additives has also proven problematic. Commercialisation and increasingly wide application of a diversity of formulations of PVC plastics since the 1930s substantially preceded both contemporary awareness of toxicity and responsive regulatory controls. As with the widespread use of lead in paint and petrol, the use of asbestos particularly in buildings, radioactive matter illuminating watch dials, and use of pesticides initially deemed beneficial, society has subsequently learned about unintended consequences and changed practices. The same is true of the many additives used in both plastic and non-plastic materials at different stages of their life cycles to provide or maintain function, as discussed previously in this book. This includes the use of various substances as PVC additives. Early metal stabilisers included cadmium and lead, both now phased out in Europe though lead stabilisation of PVC still occurs in several other parts of the world, such as in Asia and in Africa, at the time of writing, notwithstanding current knowledge about the adverse impacts of these metals entering the body largely through breakdown of products.

A problem does arise from the use of some additives due to the long service life of PVC products, as end-of-life PVC material (for example in the form of window profiles) may enter recycling streams bearing 'legacy' additives that were permissible at the time of manufacture but that have subsequently been voluntarily removed or no longer permitted in virgin PVC. We will return to the handling of legacy materials later in this book.

Other substances, such as the plasticiser DEHP (Di (2-ethylhexyl) phthalate), have been revealed as problematic and also subsequently, like lead in vehicle fuels, have been banned or voluntarily phased out in Europe and other regions in most applications. The PVC sector, particularly in Europe, has taken a leadership role in exploring and addressing these challenges on a voluntary basis, often in advance of regulatory control, as part of a wider commitment to sustainable development, and will continue so to do.

Whilst the durability of PVC is a positive asset during product life, enabling PVC products to deliver long service lives with minimal or no maintenance inputs, a third factor raising concern relates to the resilience of PVC beyond end-of-use of products. This issue relates generically to plastics of different types, as indeed to other resistant materials such as some metals, if recovery and recycling are not undertaken. The accumulation of plastic waste in the environment received high public and political prominence in the light of the 'Attenborough Effect'. Images of floating marine litter, its accumulation and damaging impacts on wildlife, in television coverage led by the veteran broadcaster Sir David Attenborough were graphic and harrowing. This triggered near-global public alarm, driving both political responses, including bans on certain types of single-use plastic applications, as well as elective phase-outs of their use by many foresighted businesses. The European Commission is addressing the accumulation of plastics in marine and wider environments through a ban on the marketing across EU Member States of throwaway plastic products for which alternatives exist.[6] Banned products include single-use plastic plates, cutlery, straws, balloon sticks and cotton buds, including cups, food and beverage containers made of expanded polystyrene and all products made of oxo-degradable plastic. However, accumulation of plastic waste remains a significant issue requiring attention and concerted global action if substantial flows of plastic into the environment, both marine and elsewhere, are to be halted. Whilst it should be pointed out that PVC is not used in the single-use applications that now comprise 80% of marine litter, and that PVC is not therefore part of this graphic problematic flotsam (in fact PVC has a density greater than that of water so would sink), addressing cyclic use with recovery and recycling remains a priority for all durable materials, including all plastics as well as substances such as aluminium. This is not merely as all polymer types get tarnished with the same simplistic brush as 'plastic pollution' in the media, and consequently in the public mind, but also as there are related issues including the potential breakdown of plastic waste into microplastics (plastic particles smaller than 5 mm in diameter) now present in every part of the environment and potentially posing risks if pollution continues at its current rate.[7]

Wider concerns about PVC were raised by the European Commission in 1997 during development of the End of Life Vehicles Directive,[8] which addressed the presence of persistent materials such as chromium and lead but also listing PVC despite it being an inert, recoverable and recyclable plastic. This led to the production of a Green Paper on *Environmental Issues of PVC* by the European Commission in 2000,[9] which concluded that "contamination of the environment by lead and cadmium should be avoided as much as possible", raised a question about the "use of phthalates as plasticisers" and recommended "that recycling of PVC should be increased". The Green Paper concluded that concerns regarding the impact of PVC on the environment "mostly related to the use of certain additives and to the management of PVC waste", calling for effective management to reduce impacts on human health and the environment of PVC throughout its life cycle.

These issues of concern are part of mounting awareness driving thinking and legislation towards a 'circular economy', seeking to economically recover and maximise material and value at the end-of-life of products. As we shall see later in this chapter, there have been substantial investments by the European PVC industry towards reducing or eliminating fugitive emissions during manufacture, shrinking the carbon/energy footprint, increasing controlled-loop life cycles for this plastic, as well as substantial attention given to the sustainable use of additives amongst other factors.

The US Environmental Protection Agency (EPA) has also paid significant attention to PVC manufacture, an example of which is concern about emissions of substances including VCM that are controlled by regulations[10] as well as some additives. Progress has also been made with clean, contained manufacture in the US. However, substantial work is still required not only in Europe, driven significantly by voluntary commitments to sustainability challenges, but around the world where less diligence and circular life cycles currently occur.

4.3 PRESSURE ON THE EUROPEAN AND UK PVC SECTOR AND THE EARLY RESPONSES

There is no doubt that PVC, at least in Europe and particularly the UK, has become an embattled industry sector with intense anti-PVC sentiment and campaigning particularly in the 1990s. Environmental pressure group activism against PVC was significant in various European nations, including the UK, in raising issues associated with chlorinated substances, spurring regulatory interest and responses both nationally and internationally. Protestor actions included break-ins at European chlorine plants in the 1980s. For all that, little traction was gained amongst the wider public until, in the late 1980s and early 1990s, pressure group tactics became more focussed on specific chlorine-containing substances. One-third of the industrial use of chlorine across the world is for the manufacture of PVC, so it was towards this widespread plastic that campaigning pressure was diverted.

Through the early 1990s, this intense campaigning pressure against PVC became a 'political football' in Denmark and Sweden. Later, the pressure group Greenpeace switched tactics in the UK to focus on retailers selling PVC-containing products rather than campaigning directly against the producers of the plastic, staging a series of demonstrations outside UK retail stores. The result was to progressively change public opinion against PVC, reinforced by press releases and the 1996 publication *Saving Our Skins*[11] followed in 1998 by *Alternatives to PVC*.[12] It should be noted that, as one example, the recommended alternative to PVC for drainage pipes and guttering in *Alternatives to PVC* was copper, a metal not only in increasingly short supply but also toxic to aquatic life and prone to bending thereby shortening its useful life; a level playing field of evaluation based on sustainability principles was clearly not observed in suggesting it as a more sustainable alternative. Fearing impacts on sales, the targeted major retailers began to take notice of these very public protests. Some manufacturers and retailers announced a phase-out of PVC with no other basis to their decision than shifting public perception, and little or no consideration of the sustainability virtues of alternative materials other than simply that they were not PVC. Some of these retailers made entirely unfounded claims that these corporate decisions were in pursuit of sustainable development, a claim for which parallel assessment of materials in common terms would have been required.

As one personal anecdote, I was invited to make a presentation on sustainable development to the Board of a leading global sportwear manufacturer. On arriving in the Boardroom in continental Europe, one member of the assembled senior management team proudly announced that the company had just taken a decision to phase out PVC. Clearly, it was expected that I should be overjoyed on receiving this news. Instead, I asked the simple question "OK, so what have you replaced it with?" The answer was a wall of baffled silence, exposing a gulf in understanding about why this question mattered. The much-trumpeted substitution of PVC, perhaps undertaken with the best of intentions but also with a complete lack of clear thinking, was in reality a ploy to surrender to media and NGO attention. However, from the perspective of sustainable development, what materials replaced PVC and how they were evaluated was of huge importance. Where was the level playing field of scrutiny that would ensure that the company was not simply investing in tomorrow's problems that were yet to be investigated and exposed? It turned out that the company simply specified technical performance of products in terms such as durability, appearance, breathability and other factors – now including no PVC – leaving it up to suppliers to decide what materials to choose in manufacturing branded products. Without betraying confidences, let us just say that we had a robust conversation about the wisdom of staking sustainability claims on allowing suppliers to choose constituent materials with no critical and scientific evaluation!

A group of UK major retailers recognised that there was an issue to manage. These retailers were mindful of the wisdom and possible irresponsibility, not to mention the expense, of abandoning PVC. There were also questions about what other materials pressure groups might next target if PVC had been replaced. Consequently, this group of major UK retailers agreed to work together to tackle this issue, establishing a *PVC Retailer Working Group* in September 1996. The *PVC Retailer Working Group* comprised the Asda, Tesco, Co-op, Waitrose and Body Shop retail chains, facilitated by the environmental NGO Greenpeace but excluding PVC manufacturers. The work of this Working Group concluded that the scientific evidence was not robust enough to justify phasing out PVC. The Group then commissioned the National Centre for Business and Ecology (NCBE) to undertake a review of evidence concerning the impact of PVC on human health and the environment, focusing particularly on packaging and construction products. The NCBE study found that, on balance of probability, there was no compelling scientific reason for retailers not to continue to use PVC.[13] In summary, the NCBE report found that there was no overall scientific reason why retailers should not continue to use PVC if "…careful manufacture, use, recycling and final disposal of PVC products to the highest standards can control the risks associated with the material to acceptable levels but will not completely eradicate them". Greenpeace, unhappy with these conclusions, left the *PVC Retailer Working Group*, leading to its reconstitution.

4.4 INDUSTRY ENGAGEMENT WITH SUSTAINABLE DEVELOPMENT

From 1998, the reformed *PVC Coordination Group* included representatives from the two UK PVC manufacturers operating at that time – EVC and Hydro Polymers – as well as the Environment Agency as it was seen as important for industry and its regulator to respond to emerging issues relating to the PVC value chain. The *PVC Coordination Group* was independently chaired by Jonathon Porritt, a prominent sustainability advocate who, at that time, was co-founder and Executive Director of the sustainable development charity *Forum for the Future*.

Woken by NGO pressure, the PVC industry had begun to recognise and address some legitimate criticism and to take steps to direct efforts towards aspects of poor environmental performance. Quality Management Systems and Environmental Management Systems were introduced. Hydro Polymers attained ISO14001 accreditation in 1996, ISO9001 accreditation in 2000 and became EMAS-compliant by 1998. EVC followed suit, attaining ISO14001 accreditation around the same time. (Both Hydro Polymers and EVC have since been acquired and integrated into INEOS Inovyn.) The industry also commissioned several Life Cycle Assessment (LCA) studies for a range of PVC applications, possibly more so than for any other materials. These LCA studies also, by majority, found that PVC was no more environmentally unacceptable or unsustainable than alternative materials when evaluated over whole life cycles. Some environmental and ethical issues were highlighted for attention by the industry.

Some broadly parallel conclusions were drawn in Europe in roughly the same time horizon. Spurred by public activism and pressure from NGOs and some Member States, the European Commission took increasing interest in PVC, commissioning the study *Life Cycle Assessment of PVC and of Principal Competing Materials* resulting in an exhaustive 325-page report in 2004.[14] The overall goal of the study was to "...compile an overview of the publicly available information on Life Cycle Assessments (LCA) on PVC and competing materials, for a variety of applications, in order to assess existing information and to identify information gaps". Aside from finding that the outcomes of LCA studies were highly sensitive to their founding goals and scope, the study concluded that raw material extraction (crude oil and rock salt) and VCM production, as well as the production of stabilisers and plasticisers, played a major role in the overall environmental impacts of PVC compounds. It was in the building and construction sector that PVC was found to offer the strongest performance over LCA criteria. For window profiles, one of the most important PVC applications, comparison of PVC and other materials revealed that no materials had an overall advantage across the standard impact categories used in LCA studies. In flooring, linoleum was found to have either comparable or slightly fewer environmental impacts to PVC flooring in the production phase, with wooden flooring tending to have lower impacts than PVC and linoleum in the production phase but being more demanding in the use and maintenance phase. In roofing applications, some polymer solutions tended to have lower environmental impacts than competitive systems. Conclusions about comparative studies of materials used for pipes were heterogeneous, though cast iron performed least well. The report concluded that economically feasible options existed at that time for the recycling of recovered PVC, recognising in particular the large amount of waste medical products generated by hospitals, with a high level of expectation of identifying methods for environmental improvement. Mechanical recycling was considered to be potentially economically feasible, though at the time was only accounted for by a small amount of post-consumer PVC waste, with emerging chemical recycling also feasible, though recycling overall was hampered by the fact that markets for recycled PVC were not yet adequately established compared, for example, with recycled metals. Overall, the European Commission report found that the environmental profiles of PVC products were broadly similar than those for other substances in equivalent durable applications, with significant opportunity for further improvements (many of which have occurred in the intervening two decades).

Nonetheless, in the face of these independent science-based evaluations, a flow of anti-PVC sentiments continued from various EU Member States, especially Denmark and Sweden.

This led the industry to take a more proactive, solutions-seeking stance from 1999 under the newly-constituted *PVC Coordination Group*. This new group commissioned NCBE to prepare an *Environmental Charter for UK PVC Manufacturers*[15] and an *Eco-efficiency Code of Practice for the Manufacture of PVC*[16] for the industry. This Charter and Code of Practice were agreed and implemented by the then two UK PVC manufacturers with a commitment to monitoring and production of independently audited reports, not only demonstrating practical commitment but also driving tangible progress on targeted eco-efficiency issues in the manufacture of PVC.

Of course, sustainable development entails a far greater scope of interdependent factors than eco-efficiency and the manufacturing phase alone. This reality led Jonathon Porritt, the independent Chair, to urge the *PVC Coordination Group* to propose that a 'sustainability gap analysis' was necessary. The *PVC Coordination Group* subsequently commissioned this sustainability gap analysis study from the UK office of international sustainable development charity The Natural Step (TNS), using the science-based TNS tools to address factors relevant to the production, use and after-use of PVC.

This is where I entered the PVC story directly, in my role at that time as Director of Science of the UK office of TNS and lead investigator of this study. Credit must be given to visionaries in the industry who realised that the woes they were facing were strategically important sustainability problems. To put some of this into TNS language, issues such as mercury releases from some chlorine production methods still active at the time, fugitive emissions of potentially problematic chemicals during manufacture, the use of some problematic heavy metal and organic substances in additives, and issues relating to the linear life cycle leading to problems stemming from cavalier disposal were already 'hitting the walls of the funnel'. A strategic vision was needed.

At a highly confidential meeting initiating this study, held at the Institute of Directors in London, Jonathon Porritt and I met with the Chief Executives and key senior sustainability staff of Hydro Polymers and EVC to scope out the 'sustainability gap analysis' study. This revolved around three questions:

- **Firstly, is the PVC industry currently sustainable?** Clearly not, given the issues forcing this meeting, but also as we live in a deeply unsustainable world in which no sector can realistically make a claim of full sustainability.
- **Secondly, if not, is it moving in the direction of increased sustainability?** Well, yes and no, acknowledging progress with prior working groups and the commitments of the UK PVC manufacturers to the *Environmental Charter for UK PVC Manufacturers* and the *Eco-efficiency Code of Practice for the Manufacture of PVC*. However, a great deal more clearly had to be done.
- **Thirdly, and most significantly, what steps must it take to become sustainable?** It is this question that the TNS Framework (the FSSD) was ideally structured to address.

Shortly before exiting this confidential meeting, I reflected that, if we got this right, I could envisage a longer-term future in which PVC had some identifiable certification marque akin to the established *Forestry Stewardship*

Council (FSC) scheme, effective since 1994 for assuring forest products, and the *Marine Stewardship Council* (MSC), instituted in 1999 to address commercial fisheries value chains. The looks on the Chief Executives' faces suggested they thought I was entirely mad or, to put it in kinder terms that this was highly unlikely!

To cut short a long story, entailing extensive networking and consensus-building exercises and detailed analysis of many facets of the PVC life cycle, 2000 saw publication of the TNS report *PVC: An Evaluation Using the Natural Step Framework.*[17] To make the key issues to be addressed more tractable for the industry, at a higher summary level than the detailed analysis using the TNS System Conditions and backcasting approach, we decided to summarise five key *TNS Sustainability Challenges for PVC*. These five TNS Sustainability Challenges for PVC are reproduced in Table 4.1.

To say that 'selling' these challenges, albeit informed by backcasting from science-based principles of sustainability, was itself challenging in the prevalent operating environment of 2000 is to underplay the reality! After all:

1. Today, the language of climate change and decarbonisation is widely used. As one example, under the European Green Deal, there is a commitment for at least 55% reduction in greenhouse gas emissions by 2030 relative to a 1990 baseline, and the attainment of climate neutrality by 2050.[18] However, in 2000 it was barely mentioned in business and regulation and still regarded with a degree of scepticism in the science community beyond a field of climate experts. This was indeed a serious challenge to a carbon-intensive industry in terms of the energy-intensity of manufacturing and the carbon embedded in its products.

TABLE 4.1   The five TNS Sustainability Challenges for PVC

1. The industry should commit itself long term to becoming carbon-neutral
2. The industry should commit itself long term to a controlled-loop system of PVC waste
3. The industry should commit itself long term to ensuring that releases of persistent organic compounds from the whole life cycle do not result in systemic increases in concentration in nature
4. The industry should review the use of all additives consistent with attaining full sustainability, and especially commit to phasing out long-term substances that can accumulate in nature or where there is reasonable doubt regarding toxic effects
5. The industry should commit to the raising of awareness about sustainable development across the industry, and the inclusion of all participants in its achievement.

2. Like virtually all industries framed by then-prevalent assumptions and regulatory framework, there was profit to be made in sales but little liability from linear product use leading to disposal. All risks beyond the factory gate were perceived as the problem of other societal sectors. Although concepts such as 'Product Stewardship' and 'Extended Producer Responsibility' were entering the lexicon, describing long-term solutions to management of waste products by shifting the responsibility for collection, transportation and management of products away from local authorities and onto manufacturers, mainstream penetration of the principles was not extensive, and markets for recycled plastics were in their infancy. Product Stewardship in the form of the 'Responsible Care' programme was launched in 1985 by the Chemistry Industry Association of Canada (formerly the Canadian Chemical Producers' Association: CCPA). Responsible Care has since grown into a voluntary initiative developed autonomously by the chemical industry on a global basis, today covering 67 countries with combined chemical industries accounting for nearly 90% of global chemical production.[19] Responsible Care focuses on health, safety, and environmental performance, with signatory chemical companies agreeing to improve performances in environmental protection, occupational safety and health protection, plant safety, product stewardship and logistics and dialogue with neighbours and the public. As a voluntary commitment, Responsible Care is independent from legal requirements but complementary to government regulation. Critics argue that, without explicit sanctions, reliance on industry self-regulation can lead, and has led to, opportunistic behaviours against the spirit of the agreement.[20] Other critics suggest that it can be more directed at influencing public concerns and opinions, whilst also opposing support for stronger and more expensive public health and environmental legislation and regulation of chemical products.[21] However, France now has some of the most progressive extended producer responsibility legislation in Europe, with waste prevention and management regulated by the French Environmental Code (FEC), as modified by the Law on the Circular economy (Law No. 2020-105 of 10 February 2020: the 'Circular Economy Law'), aimed at promoting a circular economic model based on the ecodesign of products, responsible consumption, the extension of shelf-life, reuse of products and the recycling of waste.[22] Statutory drivers towards circularity is intensifying, as for example the 2022 proposal by the European Commission for a new *Ecodesign for Sustainable Products Regulation*,[23] building on the existing *Ecodesign Directive*,[24] and

the European Commission's 2022 *Construction Product Regulation (CPR)*[25] that emphasises durability, recoverability and recyclability.

3. The PVC industry at UK and European scales had at least already launched itself on the pathway of addressing releases of persistent organic compounds from the manufacturing phases, for example under the *Environmental Charter for UK PVC Manufacturers* and the *Eco-efficiency Code of Practice for the Manufacture of PVC*. Nonetheless, extending this concept to whole product life cycles represented quite a conceptual stretch.

4. The European industry had also made progress concerning review of PVC additives, leading to reduction and phase-out of some known problematic substances. This new challenge signalled a more wholesale review of all additives and, importantly, shifted thinking towards their sustainable use throughout product life rather than considering their intrinsic chemical properties in isolation as required by then prevalent but still current legislation.

5. As noted under Challenge 2, thinking and acting at the scale of the whole life cycles of PVC products were in their infancy. However, there was a clear need to raise awareness about and integrate efforts around sustainable development across whole value chains if the prior sustainability challenges were to be progressed.

The *PVC Coordination Group* dissolved in 2002 once the retailers were assured that PVC had a viable future and the attention of pressure groups had moved on to other matters. However, this was more than a 'tick box' exercise for the more engaged players in the UK PVC industry.

There was a real risk that, with NGO and media attention now diverted, some retailers might simply then resort to buying at the cheapest price, ignoring the progress and investments made in sustainable progress. To the less aware or diligent observers, two PVC compounds might appear to be identical in characteristics, yet could have been formed from widely differing supply chains, manufacturing processes and additive types, and with differing potential for recyclability. In essence, the sustainability footprints of these two contrasting yet ostensibly similar PVC compounds could differ vastly. It was therefore necessary to develop a means to audit the sustainability credentials of different PVC products and life cycles if retailer claims of a commitment to sustainability were to be scientifically and morally defensible, for the manufacturing industries to safeguard their investments in best and improving practice, and to attract buyers seeking responsible sourcing of products differentiated by sustainability footprint.

These considerations led to the establishment of a new group in 2004, the *PVC Stakeholder Forum for Sustainability*, largely comprising members of the former *PVC Coordination Group*. The primary purposes of the *PVC*

Stakeholder Forum for Sustainability were to seek means to assure the credentials of PVC from the point of view of sustainability performance, to drive awareness of sustainability issues across PVC value chains, and to grow practical engagement across the PVC retailing, using and producing industries. The *PVC Stakeholder Forum for Sustainability* progressively evolved into a formal advisory panel influencing the UK PVC industry's decision-making processes, developing wider networks with sectors with interests in PVC, and driving further progress against the five TNS Sustainability Challenges for PVC with the aim of 'future-proofing' the sector against sustainability-relevant issues likely to emerge in future.

The UK-based PVC manufacturing company Hydro Polymers was bold in its leadership stance with respect to sustainability, in part recognising the existential threat it had faced through former campaigning pressure. Significant elements of former criticism were acknowledged as justified on the basis of previous performance. However, proactive engagement with sustainable development challenges was also driven by the values of Norsk Hydro, the Norwegian parent company of Hydro Polymers.

Hydro Polymers senior management engaged strongly with the five TNS Sustainability Challenges for PVC, including the launch a *PVC for Tomorrow* a programme bringing more employees into the process. Dr Jason Leadbitter was particularly prominent in leading Hydro Polymers' engagement with sustainable development, publishing in 2002 his own scientific paper *PVC and sustainability*[26] and inviting employees engaged in the *PVC for Tomorrow* programme to submit their best ideas about how to progress the five TNS Sustainability Challenges for PVC. The commitment was strategic and visionary, a wiser pathway than simply waiting for a renewed onslaught from campaigning environmental NGOs. This was backed up by substantial funds, ring-fenced by Norsk Hydro to support novel sustainability initiatives offering the best economic returns on investment. Some of the many initiatives implemented by Hydro Polymers in putting the five TNS Sustainability Challenges for PVC into action are summarised in Table 4.2.

These and many more breakthroughs, some substantial and subsequently influential on wider industry practices, are documented in my 2008 book *PVC: Reaching for Sustainability*.[28] The book also includes chapters exploring in detail the sustainability profiles of all major additive families assessed by TNS System Condition, and still stands as a useful resource for interested readers (Figure 4.2).

Stakeholder workshops involving both the supply chain and customers, industry-scale sharing of ideas and offerings of sustainability training were part of a wider package of initiatives by which Hydro Polymers was to prove highly influential in embedding sustainability awareness and commitment within the PVC industry.

TABLE 4.2 The five TNS Sustainability Challenges for PVC and examples of early progress by Hydro Polymers

THE TNS SUSTAINABILITY CHALLENGES FOR PVC	EXAMPLES OF EARLY PROGRESS BY HYDRO POLYMERS (DOCUMENTED IN EVERARD, 2008)
1. The industry should commit itself long term to becoming carbon-neutral	Innovation of an adiabatic cracker, implementation of a heat exchanger on a slurry stripper, modification of stores lighting, updating the road transport fleet, and installing larger silos on sites to accept bigger loads (easing scheduling and off-loading which increased vehicle payloads). Many of these innovations had payback periods of less than one year.
2. The industry should commit itself long term to a controlled-loop system of PVC waste	Promotion of recovery and recycling of post-manufacturing and post-consumer PVC to create new 'greener' PVC compound. Undertaking research to ascertain retention and functionality of additives in recyclate including, as examples, calcium carbonate and titanium dioxide.[27] Innovation of a recycled EcoVin® product with enhanced manufacturing and life cycle environmental credentials.
3. The industry should commit itself long term to ensuring that releases of persistent organic compounds from the whole life cycle do not result in systemic increases in concentration in nature	Innovation of novel seal material in reactors to avert catalytic formation of dioxins during PVC production, as well as substantial progress towards elimination of fugitive emissions.
4. The industry should review the use of all additives consistent with attaining full sustainability, and especially commit to phasing out long-term substances that can accumulate in nature or where there is reasonable doubt regarding toxic effects	Under the Vinyl2010 voluntary commitment (discussed in the following chapter), the European PVC industry committed to phase out lead from PVC compounds by 2015. Hydro Polymers had already taken a unilateral decision to do so by the end of 2007. Recycling experiments carried out by Hydro Polymers, window manufacturers and universities demonstrated that calcium carbonate filler is retained in an adequately functional state in PVC compound recycled through multiple thermal cycles.

(Continued)

TABLE 4.2 (Continued)   The five TNS Sustainability Challenges for PVC and examples of early progress by Hydro Polymers

THE TNS SUSTAINABILITY CHALLENGES FOR PVC	EXAMPLES OF EARLY PROGRESS BY HYDRO POLYMERS (DOCUMENTED IN EVERARD, 2008)
5. The industry should commit to the raising of awareness about sustainable development across the industry, and the inclusion of all participants in its achievement	The *PVC for Tomorrow* programme involved about 140 of the 1,400 total employees of Hydro Polymers in TNS-based education, also instituting a competition across Hydro Polymers sites to present their best ideas concerning addressing the five TNS Sustainability Challenges for PVC. Hydro Polymers set out to bring its strategic additive suppliers into the quest for sustainability for PVC, including through annual sustainability conferences, as well as strategic dialogue with some customers.
Synergies across the five TNS Sustainability Challenges for PVC	All five TNS Sustainability Challenges for PVC are systemically interconnected. Therefore, as a minimum, progress with one challenge must not inhibit progress with others. However, in practice, innovations can contribute positively to multiple challenges. For example, selection of additives that are inherently recyclable can promote progress with controlled-loop management, with reduced energy usage, retention of embodied energy and associated carbon emissions, and with little or no potential for problematic fugitive emissions. This helps other players along PVC value chains appreciate the benefits of proactive engagement with sustainable development.

FIGURE 4.2   The cover of Mark Everard's 2008 book *PVC: reaching for sustainability* ©Mark Everard.

4.5 BREAKING THROUGH THE PREJUDGEMENTS

Society has progressed massively in addressing environmental and social ills through the efforts of campaigning NGOs, focusing emergent societal unease and working with media to advocate measures to recognise and address major problems. The influence of NGOs in the unfolding of the 'environment movement' has elevated the political profile of issues as diverse as ozone depletion, animal welfare, conservation of wild species and fisheries, child and indentured labour, climate change, whaling and pesticide accumulation. NGOs have been key change agents, harnessing and focusing societal concerns to drive legislative, market, consumer behaviour and other changes.

The significant attention that PVC received from campaigning environmental NGOs from the 1980s and 1990s, especially in Europe, was influential on the thinking of governments and retailers. There is some suspicion that this may have been implicitly or explicitly supported by some competitive industries, for example in the case of advertisements for zero-chlorine cables generating 'acid-free fumes' when burned, yet omitting to mention that PVC is self-quenching whilst other more flammable alternative materials require flame-retardant additives with their own associated problems. It is nonetheless true that various aspects of PVC manufacturing practices and supply chains, many of them historically in Europe but some still prevalent overseas, as well as the chemistry of some problematic additive substances, had created a case to answer. Whilst the supply chains, manufacturing practices and level of concern about the fate of product beyond the factory gate were far from blameless in other industry sectors at that time, leading players in the European PVC sector have been, with hindsight, happy to acknowledge that this pressure caused the industry to reflect upon and institute improvements to their practices. This has had the longer-term benefit of causing it to recognise the importance of engaging proactively with sustainable development as a uniting framework that will be increasingly influential in shaping future markets and societal 'licence to operate'.

What is less helpful is that, once sustainable development began to be grasped as a strategic challenge by the European PVC sector, the actions and communications of some pressure groups failed to recognise progress and continued to promote an anachronistic view on issues that, though perhaps still prevalent in other territories, had already been phased out or dealt with across Europe. Concerted NGO and lazy media repetition have solidified this historic picture in some minds as representing current practice.

As noted earlier in this book, branding any material as inherently 'bad' can lead to the unintended negative consequence of uncritical knee-jerk replacement with other materials that may not have been critically assessed on the same terms, resulting in naïve substitution decisions that risk locking investment into tomorrow's as yet untested problems. The reality is that PVC is just one amongst a huge diversity of natural and synthetic substances used by society sharing common sustainability challenges, known, assumed or yet to reveal themselves if they were subject to equivalent critical risk analysis across whole societal product life cycles.

As addressed in some detail previously, problems arise when we base judgements solely on intrinsic chemical properties rather than overall life cycle risk issues relating to actual exposure. Also, critically, many former approaches have entirely ignored the 'lost half' of sustainable development in terms of recognising the benefits that use of materials confers for the meeting of human and environmental needs in the safest and most efficient manner. To achieve this, it is essential to take account of such properties as durability and long service life, resilience and low maintenance inputs, and both theoretical recyclability and actual operational or planned recycling systems for material and value recovery at end-of-life.

That all said, sometimes we will indeed need to swap the materials we use with more benign ones. In the case of PVC, a short-life single use in food delivery might result in unrecoverable and contaminated waste that is likely to accumulate in nature if incautiously thrown away. We see this in particular for polythene and other durable polyolefin materials that are used for these purposes, and which are the major constituents accumulating as gyres of plastic waste in marine systems. Issues with single-use products and materials are, of course, not just related to plastics, also relevant to the disposal and environmental accumulation of resistant materials such as metals (for example aluminium, steel and iron) and coated paper but also liquids such as oils and surfactants. A strategic challenge for society is to break industrial practices and the antisocial or careless behaviours of citizens that perpetuate the cavalier discarding of used products into the environment.

Nonetheless, some single-use applications are entirely justified within controlled life cycles. This is particularly the case for various medical applications, for which the properties of PVC make very significant and overriding contributions to meeting human needs, particularly for infection control purposes. Recovery and incineration at temperatures that break down potentially hazardous combustion products are well established. Better still, across Europe, a substantial bulk of PVC medical devices is now mechanically recycled into new technical applications, for example under the medical device recycling scheme PVCMed Alliance[29] operating at several large European hospitals, particularly in Belgium. This programme was set up in recognition that PVC is the single most used polymer for disposable medical equipment, such as oxygen and anaesthetic masks, tubing, IV and dialysis bags, recognising that the plastic is also easily recyclable, retaining its technical properties through numerous recycling cycles. Increasing numbers of hospitals are joining the PVCMed Alliance programme, under which technical experts also teach staff about correct sorting for recycling. In the UK, the RecoMed programme has been established since 2016 to facilitate the recycling of the estimated 7 tonnes of PVC masks and tubing used each year by NHS hospitals that could be available for recycling (2015 figure), putting in place collection infrastructure, sorting and logistics for single-use medical devices to be recycled into potentially recyclable vinyl flooring, and offering regular updates to participating hospitals on the amount of material processed.[30]

For many other long-life applications, particularly in building and construction (accounting for 69% of total PVC production) as well as other forms of civil infrastructure, the durability, low or no maintenance requirements, and recyclability of PVC means that it is ideally suited to meeting human needs efficiently and safely with the potential for recovery and reuse at the end of long service life. Where less durable substances are used in these longer-life applications, they tend to offer shorter effective lives and hence a lower level of meeting of needs and, to prolong service life, tend also to require substantial maintenance inputs and replacement with their additional sustainability footprints.

The clarion cry throughout this book is for a level playing field of assessment. PVC is not a material apart from all others. As with many divisive issues, polarisation has often blocked progressive and scientifically informed action. Over-simplistic debate about 'good' versus 'bad' materials has distracted us from thinking systemically, taking account of broader dimensions of sustainability including how best to meet human needs in the safest and most efficient manner when full product life cycles are factored into decision-making. At this pivotal time in human history, framed by pressing biodiversity, climate change and other linked sustainability challenges, what we now require is the good sense and sophistication to think in terms of how best we can use and innovate materials, whatever that material type, to best serve human needs in the safest and most efficient manner possible.

We also have to take a strategic view, for which we need clear navigation to inform us about how the incremental steps that can be taken in the short term can be aligned towards the eventual attainment of ultimate sustainability goals that may not be immediately or rapidly attainable. In our far less than sustainable world, this means that we cannot reasonably expect to make an instant leap to perfection. I am often minded of the aphorism of the French Enlightenment writer, historian and philosopher Voltaire (1694–1778), paraphrased here as not to let the best become the enemy of the good. If the 'best' is full sustainability, then a 'good' decision could be a material use choice or innovation that is a wise and currently attainable step on an incremental journey towards the goal of full sustainability, founded on assessment against a systemically framed 'level playing field'. This is so for material manufacturers and their supply chains, and must become so for the assessment criteria used by those specifying and purchasing products, and planning end-of-life recovery and recycling. It must also become a foundation for evolution in legislation and supporting tools if they are to propel society towards the goal of sustainability.

NOTES

1 etymonline.com. (n.d.). *vinyl (n.).* etymonline.com. [Online.] https://www.etymonline.com/word/vinyl, accessed 11 April 2023.

2 KIWA. (2022). *Leak Tightness of PVC Fittings with Hydrogen.* Kiwa Technology B.V., Apeldoorn. [Online.] https://www.netbeheernederland.nl/_upload/Files/Rapport_Leak_tightness_of_PVC_fittings_with_hydrogen_254.pdf, accessed 13 May 2023.

3 Geyer, R., Jambeck, J.R. and Levender, K. (2017). Production, use, and fate of all plastics ever made. *Science Advances*, 3(7), e1700782. DOI: https://www.science.org/doi/epdf/10.1126/sciadv.1700782.

4 OPCW. (2023). *Chemical Weapons Convention*. Organisation for the Prohibition of Chemical Weapons (OPCW). [Online.] https://www.opcw.org/chemical-weapons-convention, accessed 13 April 2023.

5 Shaheen, K. (2015). Assad regime accused of 35 chlorine attacks since mid-March. *The Guardian*, 24 May 2015. [Online.] https://www.theguardian.com/world/2015/may/24/syria-regime-accused-of-using-chlorine-bombs-on-civilians, accessed 13 April 2023.

6 European Commission, Directorate-General for Environment. (2021). *Turning the Tide on Single-Use Plastics*. Publications Office, European Commission. [Online.] https://data.europa.eu/doi/10.2779/800074, accessed 12 April 2023.

7 SAPEA. (2019). *A Scientific Perspective on Microplastics in Nature and Society*. Scientific Advice for Policy by European Academies (SAPEA), 15 January 2019. [Online.] https://sapea.info/topic/microplastics/, accessed 13 April 2023.

8 CEC. (1997). *Proposal for a COUNCIL DIRECTIVE on End of Life Vehicles, COM(97) 358 final 97/0194 (SYN)*. Commission of the European Communities (CEC), Brussels. [Online.] https://eur-lex.europa.eu/LexUriServ/LexUriServ.do?uri=COM:1997:0358:FIN:EN:PDF, accessed 13 April 2023.

9 CEC. (2000). *Green Paper: Environmental Issues of PVC*. Commission of the European Communities (CEC), Brussels. [Online.] https://ec.europa.eu/environment/pdf/waste/pvc/en.pdf, accessed 13 April 2023.

10 EPA. (2023). *Polyvinyl Chloride and Copolymers Production: National Emission Standards for Hazardous Air Pollutants (NESHAP) - 40 CFR 63 Subparts J & HHHHHHH*. US Environmental Protection Agency (EPA). [Online.] https://www.epa.gov/stationary-sources-air-pollution/polyvinyl-chloride-and-copolymers-production-national-emission-0#rule-summary, accessed 13 April 2023.

11 Greenpeace. (1996). *Saving Our Skins*. Greenpeace.

12 Greenpeace. (1998). *Alternatives to PVC*. Greenpeace.

13 NCBE. (1997). *Summary Report for PVC Retail Working Group*. National Centre for Business and Ecology, Manchester.

14 European Commission. (2004). *Life Cycle Assessment of PVC and of Principal Competing Materials*. European Commission. [Online.] https://vdocuments.mx/pvc-final-report-lca-en.html?page=1, accessed 29 April 2023.

15 NCBE. (1999). *Environmental Charter for UK PVC Manufacturers*. National Centre for Business and Ecology, Manchester.

16 NCBE. (1999). *Eco-efficiency Code of Practice for the Manufacture of PVC*. National Centre for Business and Ecology, Manchester.

17 Everard, M., Monaghan, M. and Ray, D. (2000). *PVC: An Evaluation Using the Natural Step Framework*. The Natural Step, Cheltenham.

18 European Commission. (2023). *2030 Climate Target Plan*. European Commission, Brussels. [Online.] https://climate.ec.europa.eu/eu-action/european-green-deal/2030-climate-target-plan_en, accessed 13 May 2023.

19 ICCA. (2023). *Responsible Care®*. International Council of Chemical Associations (ICCA). [Online.] https://icca-chem.org/focus/responsible-care/, accessed 14 May 2023.

20 King, A.A. and Lenox, M.J. (2000). Industry self-regulation without sanctions: the chemical industry's responsible care program. *The Academy of Management Journal*, 43(4), pp.698–716. DOI: https://doi.org/10.2307/1556362.

21 Givel, M. (2007). Motivation of chemical industry social responsibility through Responsible Care. *Health Policy*, 81(1), pp.85–92. DOI: https://doi.org/10.1016/j.healthpol.2006.05.015.

22 CMS. (2023). *Plastics and Packaging Laws in France*. CMS Law. [Online.] https://cms.law/en/int/expert-guides/plastics-and-packaging-laws/france, accessed 13 May 2023.

23 European Commission. (2022). *Ecodesign for Sustainable Products*. European Commission, Brussels. [Online.] https://commission.europa.eu/energy-climate-change-environment/standards-tools-and-labels/products-labelling-rules-and-requirements/sustainable-products/ecodesign-sustainable-products_en, accessed 14 May 2023.

24 European Parliament. (2009). *DIRECTIVE 2009/125/EC OF THE EUROPEAN PARLIAMENT AND OF THE COUNCIL of 21 October 2009 establishing a framework for the setting of ecodesign requirements for energy-related products*. European Parliament, Brussels. [Online.] https://eur-lex.europa.eu/legal-content/EN/TXT/PDF/?uri=CELEX:02009L0125-20121204&from=EN, accessed 14 May 2023.

25 European Commission. (2022). *Construction Product Regulation (CPR)*. European Commission, Brussels. [Online.] https://single-market-economy.ec.europa.eu/sectors/construction/construction-products-regulation-cpr_en, accessed 14 May 2023.

26 Leadbitter, J. (2002). PVC and sustainability. *Progress in Polymer Science*, 27(10), pp.2197–2226. DOI: https://doi.org/10.1016/S0079-6700(02)00038-2.

27 Leadbitter, J. and Bradley, J. (1997). *Closed Loop Recycling Opportunities for PVC*. Institute of Polymer Technology and Materials Engineering, Loughborough University; 3–4 November 1997.

28 Everard, M. (2008). *PVC: Reaching for Sustainability*. IOM3 and the Natural Step, London. 269pp.

29 PVCMed Alliance. (n.d.). *Medical device recycling*. PVCMed Alliance. [Online.] https://pvcmed.org/sustainability/recycling, accessed 12 April 2023.

30 RecoMed. (2023). *How RecoMed Works*. RecoMed. [Online.] https://recomed.co.uk/how-recomed-works/, accessed 13 May 2023.

Voluntary Sustainability Commitments of the European PVC Sector

5

Sustainability, or at least environmental awareness and action, had been attracting the attention of the PVC sector more widely across Europe over the same time period as in the UK, with shared initiatives across the European sector. For example, the European Council of Vinyl Manufacturers (ECVM) established the 'ECVM Charter' in 1995, initially for S-PVC (suspension PVC). The ECVM Charter was and continues to be focused on ensuring low organochlorine emissions (below regulatory limits) from manufacturing and processing, and has been updated at intervals including, in 2001, extension to include E-PVC (emulsion PVC).[1]

In parallel with initiatives taking place in the UK, a group of four European PVC trade associations – the European Council of Vinyl Manufacturers (ECVM), the European Plastics Converters (EuPC), the European Council for Plasticisers and Intermediates (ECPI) and the European Stabiliser Producers Association (ESPA) – started to work together to develop and gain the support from members for a voluntary code of practice known as *Vinyl2010*. Vinyl2010, consolidated in 2000, was intended to become the European PVC industry's chosen voluntary route to the achievement of sustainability, putting in place a 10-year plan running to 2010.

DOI: 10.1201/9781003453949-5

5.1 VINYL2010

Central to Vinyl2010 was a *Voluntary Commitment* of the European PVC industry intended to enhance the sustainability profile of the European PVC industry by improving production processes and products, investing in technology, minimising emissions and waste and boosting collection and recycling of end-of-life PVC products. Hydro Polymers was a leader, and major funder, of Vinyl2010 with innovations within the company to progress the five TNS Sustainability Challenges for PVC feeding into the understandings of Vinyl2010 (Figure 5.1).

FIGURE 5.1   The Vinyl2010 logo. ©VinylPlus.

Real progress and partnership working occurred under Vinyl2010, though some external critics regarded it as little more than an industry-led attempt to avert more stringent European regulation. In part, this was true, as proactive engagement with issues highlighted by foresighted sustainability 'gap analysis' can flag up issues not yet on the regulatory radar. The Vinyl2010 initiative, instigated and managed from within the industry, nonetheless began to achieve tangible buy-in and some areas of progress on selected issues. However, it lacked a foundation in clear scientific sustainability principles against which its actions and outcomes could be audited. This exposed the industry to the risks of claims being undermined by pressure groups still seeking to exploit weaknesses.

5.2 VINYLPLUS® TO 2020

Beyond 2010, a new framing for the Europe-wide voluntary commitment of the PVC sector was required. After significant dialogue, the VinylPlus® programme was launched on 22 June 2011. Central to the VinylPlus voluntary commitments were adoption of the five TNS Sustainability Challenges for PVC in a slightly reworded from, represented as an interlocking set (see Figure 5.2). For each challenge, targets were set and independent auditing of progress against them was reported annually. The rewordings and reordering of the five challenges are reproduced in Table 5.1, mapped against the original TNS Sustainability Challenges for PVC.

FIGURE 5.2   The VinylPlus logo and its articulation of the five sustainability challenges. ©VinylPlus.

Growing commitment and funding mechanisms throughout the decade to 2020 saw substantial audited progress against these targets reported annual in a series of VinylPlus Annual Reports openly available on the VinylPlus website (http://www.vinylplus.eu/).

5.3 ADDRESSING THE UN SUSTAINABLE DEVELOPMENT GOALS

In 2015, midway through the first decadal phase on the VinylPlus programme, the world agreed on a new direction for development in the shape of the 17 UN Sustainable Development Goals (SDGs) and their associated 169 targets.[2] This was momentous as, though the 17 SDGs have their detractors, they represent a globally consensual set of linked outcomes to which all signatory nations have committed on a journey to 2030. Human and environmental needs and rights, economic activity, infrastructure and institutions are all included within a systemic whole (Figure 5.3).

VinylPlus was quick to link its activities to the Goals, mapping activities under each of the five sustainability challenges to targets under each of the SDGs.[3] Undoubtedly, PVC and its products as well as material management

TABLE 5.1 Rewording and ordering of the five TNS Sustainability Challenges for PVC underpinning the VinylPlus voluntary commitment

VINYLPLUS® VOLUNTARY COMMITMENT	PVC EVALUATION USING THE NATURAL STEP FRAMEWORK
Challenge 4. Sustainable energy and climate stability – We will help minimise climate impacts through reducing energy and raw material use, potentially endeavouring to switch to renewable sources and promoting sustainable innovation.	Challenge 1. The industry should commit itself long term to becoming carbon-neutral
Challenge 1. Controlled-loop management – We will work towards the more efficient use and control of PVC throughout its life cycle.	Challenge 2. The industry should commit itself long term to a controlled-loop system of PVC waste
Challenge 2. Organochlorine emissions – We will help to ensure that persistent organic compounds do not accumulate in nature and that other emissions are reduced.	Challenge 3. The industry should commit itself long term to ensuring that releases of persistent organic compounds from the whole life cycle do not result in systemic increases in concentration in nature
Challenge 3. Sustainable use of additives – We will review the use of PVC additives and move towards more sustainable additives systems.	Challenge 4. The industry should review the use of all additives consistent with attaining full sustainability, and especially commit to phasing out long term substances that can accumulate in nature or where there is reasonable doubt regarding toxic effects
Challenge 5. Sustainability awareness – We will continue to build sustainability awareness across the value chain – including stakeholders inside and outside the industry – to accelerate progress towards resolving our sustainability challenges.	Challenge 5. The industry should commit to the raising of awareness about sustainable development across the industry, and the inclusion of all participants in its achievement

and partnerships all have significant contributions to make towards global sustainability. As noted earlier in this book, consensus around the SDGs was also highly significant in resurrecting a 'lost half' of sustainable development that had been buried under a rather unambitious perception and regulatory focus on minimum performance standards. Reframing global aspirations for sustainable development on the meeting of human, social and environmental needs reasserted this 'lost half' of sustainable development[4] and, with it, the boldly

FIGURE 5.3 The 17 UN Sustainable Development Goals (SDGs). ©United Nations.

intergenerational framing of the 1987 'Brundtland definition' of sustainable development.[5] And, under that reinvigorated vision, the emphasis on the use of chemicals could shift from a dispiriting focus on their potentially negative intrinsic properties towards the contributions that their use could make to addressing human and environmental needs and progress in the safest and most environmentally efficient manner. This positive reframing is a stimulant of opportunity and innovation to address future markets inevitably shaped by ever-tightening sustainability pressures.

Towards this aim, I was able to link my action research supporting sustainable water management in testing climatic conditions in the developing world addressing the SDGs, connecting it with my work with VinylPlus. A key output of this was a 2017 article *Repurposing business around the meeting of human needs* that considered the potential contributions of PVC water pipes, including their production and management, in supporting the water management needs of people in the semi-arid region on the edge of the Thar Desert of Rajasthan, India.[6] Briefly summarising a subset of issues highlighted in Figure 5.4, better control of water can directly contribute to water management (SDG6) and also rural infrastructure (SDG11), better servicing food security (SDG2) and health needs (SDG3) whilst protecting life on land (SDG15) and life in water (SDG14). Informed by other of my research, more efficient water management has been found to have a disproportionate benefit for the equality of women (SDG5) upon whom otherwise the drudgery of water collection falls, enabling them instead to contribute actively to farming (SDG2) and healthcare (SDG3) as well as education in which girls can participate rather than accompanying mothers on risky water-fetching duties (SDG4). Responsible

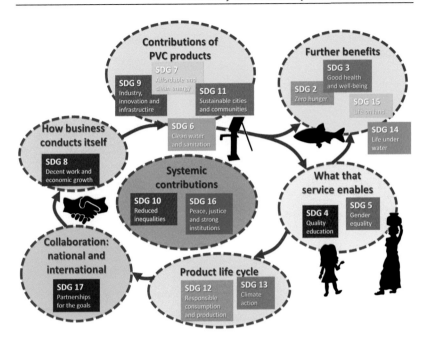

FIGURE 5.4   Potential contributions to the SDGs from water management using PVC pipes in the Thar Desert edge. ©Mark Everard.

stewardship of plastics can contribute to carbon efficiency (SDG13) and decent employment (SDG8), with meta-level coordination supporting poverty alleviation (SDG1), reduction of inequalities (SDG10) and harmonious relationships (SDG16) within a partnership making progress with the goals (SDG17).

This reconceptualisation of sustainable development towards human, environmental and civil goals emphasises the roles of chemical and material use in the meeting of needs in optimally ethical, safe and efficient means. This in turn calls for a level playing field for evaluation of chemical use – an objective, scientific and sustainability-relevant framework for comparison of which materials can achieve this most sustainably – a theme to which we will return throughout the rest of this book.

5.4 VINYLPLUS TO 2030

Beyond 2020, VinylPlus reframed its commitments into three Pathways. A line of sight can still be drawn from the scientific principles underpinning The Natural Step systems science model, the four TNS System Conditions, the five

TNS Sustainability Challenges for PVC, the five VinylPlus challenges to 2020 and the three new VinylPlus Pathways to 2030 into which they are subsumed, as illustrated in Figure 5.5.

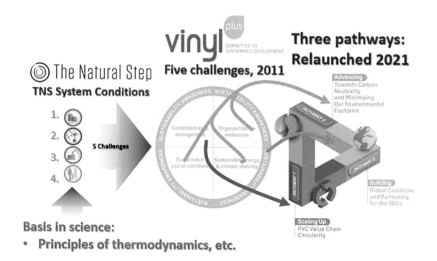

FIGURE 5.5 'Line of sight' from scientific principles to the three VinylPlus Pathways. ©Mark Everard; TNS logo ©The Natural Step.

5.5 EVOLVING VINYLPLUS ACCREDITATION SCHEMES

So how do you signify a PVC product that has been produced and has a societal life cycle consistent with a set of sustainability standards? After all, an ostensibly identical product may have been derived from raw materials sourced with a lack of environmental or social concern, manufactured without best practice, may contain additives that are potentially problematic in use and may not be ideally recyclable. Yet, to all intents, it may appear identical to the product manufactured in a responsible and auditable manner.

This was the issue that I was voicing way back in 1999 to the chief executives of the two UK PVC manufacturing companies in that room in the Institute of Directors in London, suggesting the need for an identifiable certification marque akin to the FSC or MSC in some longer-term future. Though the response at the time was bafflement, VinylPlus over the intervening years has developed a range of independently accredited certification schemes precisely to identify market-differentiated PVC products produced in accordance with its voluntary sustainability commitments.

The most visible of these independently certified schemes is the VinylPlus Product Label. The VinylPlus Product Label makes it easy for customers and markets to identify the most sustainable and high-performance PVC products. At the time of writing, the Product Label scheme covers PVC building products. Amongst the benefits of the Product Label for companies manufacturing PVC products are that it makes visible to the market that the PVC solutions are independently verified as having manufactured in accordance with sustainability principles, creating an incentive for buyers and specifiers to select these products potentially bolstering sales, and benefitting from Product Label promotion by VinylPlus. Details of the VinylPlus Product Label accreditation process are published online by VinylPlus[7] and are updated over time. They are open for application by all VinylPlus Partners, and for non-partners on payment of a Scheme Management Contribution.

The VinylPlus Product Label may be awarded for a specific PVC product, or a product family sharing generically similar characteristics. Product Label accreditation takes account of evidence for a range of criteria spanning eight categories, three of them to do with corporate performance and five specifically targeting the five sustainability challenges as listed in Table 5.2. Applicants for the VinylPlus Product Label provide a portfolio of evidence to an independent auditor, scoring a minimum level of qualifying points to achieve the award of the VinylPlus Product Label (Figure 5.6).

The goal of creating a marque for PVC-containing products compliant with VinylPlus voluntary commitments to sustainable development, distinguishing them from ostensibly similar PVC products imported or manufactured without such accredited assurance, is beginning to influence markets. Ideally, like FSC, MSC and other accreditation schemes such as the PEFC (Programme for the Endorsement of Forest Certification[8]: the world's largest forest certification

TABLE 5.2   Categories of criteria for accreditation under the VinylPlus Product Label

1. VinylPlus partnership and programme support	Responsible corporate performance
2. Organisational management requirements	
3. Supply chain management requirements	
4. Controlled-loop management and recycling	Compliance with the five VinylPlus sustainability challenges
5. PVC resin from sustainable sources	
6. Responsible use of additives	
7. Sustainable energy and climate policies	
8. Sustainability awareness and communication	

FIGURE 5.6   The VinylPlus Product Label logo. ©VinylPlus.

organisation), the VinylPlus Product Label will continue to grow in acceptance and use by specifiers pursuing responsible sourcing in public procurement and building companies, as well as amongst wider consumers seeking optimally sustainable products. One such success is explicit mention of the VinylPlus Product Label in the *Italian Official Journal* (6 August 2022, N° 183) as a valid option to prove minimal recycled content (2.2.1, p. 41), with the Product Label also mentioned as a valid reference for recycled PVC in the technical specifications of the product (2.5, p. 58).[9] The relevant text reads, "…per i prodotti in PVC, una certificazione di prodotto basata sui criteri 4.1 'Use of recycled PVC' e 4.2 'Use of PVC by-product', del marchio VinylPlus Product Label, con attestato della specifica fornitura" (translated here as "…for PVC products, a product certification based on the 4.1 criteria 'Use of recycled PVC' and 4.2 'Use of PVC by-product', of the VinylPlus Product Label, with certificate of specific supply").

For converters of PVC products, assessment of all ingredients against the various criteria under the eight VinylPlus Product Label categories entails a significant amount of work. However, to ease the burden, there is an intermediate, independently audited VinylPlus Supplier Certificate (VSC). There are two types of VSC, one for additive manufacturers and another for those manufacturing PVC compounds for onward sale to converters. Additive or compound manufacturers applying for the VSC follow an accreditation process published online by VinylPlus[10] that is periodically updated. The eight categories of VSC criteria are the same as those listed in Table 5.2, with details of the criteria adapted for these purposes. An independent VSC Auditor reviews a portfolio of evidence supplied by applicant VinylPlus members, who must score a minimum level of qualifying points to achieve the award of the VinylPlus Supplier Certificate. For additives, the VSC may apply to a single additive product, an additive product system (parent additive and a set of 'mirroring' additive products that represent slightly different variations of this parent additive product) or an additive product range (a group of additives having at least one attribute in common or that complement each other in significant ways). The VSC certificate demonstrates to the converter that the additives or compounds they are buying have been audited as compliant with VinylPlus sustainability criteria, and require no further work when applying for the VinylPlus Product Label. The VSC logo is reproduced in Figure 5.7.

VinylPlus®
Supplier
Certificates

FIGURE 5.7   The VinylPlus Supplier Certificate (VSC) logo. ©VinylPlus.

5.6 ADDITIVE SUSTAINABILITY FOOTPRINT (ASF)

PVC additives are diverse and can be complex. For this reason, the use of additives is the focus of a discrete challenge under both the TNS Sustainability Challenges for PVC (Challenge 3) and, in slightly modified form, under the VinylPlus Challenges (Challenge 4). These challenges are, as we have seen, subsumed within the VinylPlus Pathways to 2030, and are also incorporated within the criteria of the VinylPlus Product Label and the VSC. It is important to emphasise that this challenge relates to the USE of additives throughout the life cycle of the products into which they are incorporated, rather than potential hazard viewed without taking account of life cycle risk.

As discussed previously in this book, 'safe' substances such as water, oxygen and nitrogen can be risky or deadly in context. Another highly pertinent example entailing chlorine chemistry is that the human stomach contains substantial amounts of the hazardous chemical hydrogen chloride (hydrochloric acid) at a pH of 1.5–3.5 that, if encountered anywhere else in the body, would be hugely damaging. Yet it is contained, managed and serves a perfectly safe and indeed essential function in the stomach. This same principle applies for other inherently hazardous substances. If a processing aid or additive constituent is wholly consumed or converted in contained manufacturing facilities, exposure and hence risk are eliminated. Equally, a substance firmly bound into an inert compound that is then recovered at end-of-life, ideally recycled for reuse for further product life cycles but potentially safely disposed of, may be considered safe as human and environmental exposure, and hence risk, are avoided.

As noted previously, many pre-existing chemical assessment approaches focus on intrinsic chemical properties, and particularly potential hazard, in isolation from 'real world' contexts. The absence of assessment of the use of additive substances in the context both of whole product life cycles and taking account of sustainability criteria beyond narrow environmental factors necessitated the innovation of a bespoke tool to address this PVC sustainability challenge. The tool that was developed is the Additive Sustainability Footprint (ASF). ASF is adapted from tools of The Natural Step, specifically the TNS Sustainability Life Cycle Assessment (SLCA) approach.[11] Central to the ASF tool and the TNS SLCA process is the orientation of the four TNS System Conditions against the six ISO14040:2006 life cycle stages,[12] as presented in Figure 5.8. This is enacted through a ten-step process. The derivation and operation of ASF are described in greater detail in the peer-reviewed scientific paper *Additive Sustainability Footprint (ASF): rationale and pilot evaluation of a tool for assessing the sustainable use of PVC additives.*[13]

Assessment within each of these TNS System Condition/life cycle stage cells is based on a list of seven generically similar questions. Three of these are

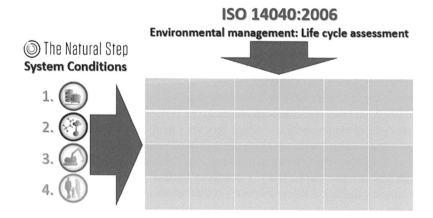

FIGURE 5.8   Representation of the orientation of TNS System Conditions against the six ISO14040 life cycle stages in the heart of the ASF approach. ©Mark Everard; TNS logo ©The Natural Step.

'impact questions', addressing negative criteria, some of which can be transferred from existing investments in REACH, LCA and other existing assessments. The further four 'progress questions' address positive benefits of the use of the additive substance and commitments in hand to future progress. Recognition of benefits is important in the context of the purpose for which additives are developed and used to better serve societal needs, a factor serially overlooked in pre-existing chemical assessment approaches. These seven generically similar questions are tailored to the specifics of each TNS System Condition/life cycle stage cell. Answers comprise three components: (1) overall response ('YES', 'NO', 'Not applicable' or 'Don't know'), (2) assessment of confidence level ('HIGH' or 'LOW') and (3) comments/references documenting supporting evidence or reasoned justification for answers provided. Examples are given in the referenced scientific paper. Although the sets of seven questions in each of the 24 cells mean that a total of 168 responses are required, research is underway to automate some answers within an online tool, and many answers are transferrable from related ASF assessments.

Overall responses to the seven questions are arithmetically converted based on the number of questions answered positively. Knowledge gaps and negative responses both score as a zero, as knowledge gaps represent 'sustainability blind spots' that may be improved with further information-gathering or targeted improvement. These totted up scores can be presented illustratively through 'traffic lights' colour coding by TNS System Condition and life cycle stage, graphically highlighting actual or potential 'unsustainability hotspots' warranting further information-gathering or innovation leading to more sustainable use of additives.

A final ASF 'snapshot report' can then be signed off by its authors and reviewers to guide further research, innovation or as evidence used as input for some criteria of the VinylPlus Supplier Certificate (VSC). The identified 'unsustainability hotspots' can guide development at relevant stages in the life cycle, for example in gathering more information on ethical dimensions of supply chains, changing material acquisition to better regulated sources, reformulation to increase product service life or lowering of maintenance requirements, marking of compound to aid sorting, or collaboration with other societal sectors to improve recovery and recycling at product end-of-life.

A scientific paper published in 2022, *Assessment of the sustainable use of chemicals on a level playing field*,[14] summarised findings of an ASF assessment of a metal-based stabiliser additive in the context of the life cycle of PVC compound used in a generic EU Window profile (frame). These findings are summarised in Table 5.3. Potential positive (☺) and negative (☹) implications are highlighted, helping identify positive contributions across life cycle stages as well as informing a 'heat map' of where innovation is required in the additive supply chain and manufacturing, during the window profile life cycle, and at and beyond end-of-life to improve overall sustainability. One of the key roles of ASF is to inform innovation relating to the use of chemicals across the societal life cycle of products. For this reason, potential relevant mitigation and innovation measures are identified (↑) at specific points in the PVC product life cycle.

The issues, benefits and items for further innovation highlighted in Table 5.3 are from a 'real world' ASF assessment by the European PVC industry. The colour coding is purely illustrative, though it does accurately highlight that most of the issues entailed in the metal used for this purpose are at the extraction phase, if not from recycled sources, with metal stabiliser additives then contributing to long service life with no inputs and hence meeting human needs to a high level, and with controlled-loop recycling then dealing with potential problems at end-of-life.

By highlighting where priority sustainability issues occur across the TNS Systems Condition and life cycle stages, ASF then informs areas for innovation across the societal life cycle of products to improve the overall sustainable use of additives and other constituent substances. This aspiration to achieve full sustainability by backcasting from principles is illustrated in Figure 5.9.

5.7 PROGRESS WITH VINYLPLUS TARGETS

Successes for PVC are summarised as progress reported by VinylPlus with the five (modified) TNS Sustainability Challenges in Table 5.4.

5.8 NOVEL PVC PRODUCTS AND LIFE CYCLES DRIVEN BY SUSTAINABLE DEVELOPMENT CHALLENGES

The purpose of backcasting from sustainability principles, and the five TNS Sustainability Challenges for PVC that flowed from the initial study and have subsequently informed VinylPlus and member company strategies, is to aid navigation towards increasingly sustainable, more desirable and less risky products and life cycles. The recognition that both products and their whole societal life cycles are important reflects the fact that the journey of sustainable development requires joining up across societal sectors, applying not just

TABLE 5.3 Potential sustainability issues by TNS System Condition associated with the use of metal-based stabiliser additives to PVC, including potential positive (☺) and negative (☹) implications, and innovation and mitigation measures (↑) at specific points in the final PVC window profile life cycle

	LIFE CYCLE STAGES				
	ADDITIVE PRODUCTION	PACKAGING AND DISTRIBUTION	COMPOUNDING/ CONVERTING	PRODUCT USE/ MAINTENANCE	POST-USE MANAGEMENT
RAW MATERIALS ACQUISITION					
TNS System Condition 1 ☹ Risk of persistent heavy metals released in mining ☺ Cadmium or lead were a significant risk, but both are now phased out from manufacturing in Europe by voluntary commitment ↑ Cease procurement from additive suppliers not signing up to European voluntary standards ↑ Substitute with light metallic elements more abundant in nature, or derive metals from recycled sources	☹ Risk of metal release in manufacturing ↑ Controlled manufacturing eliminating material losses, with post-industrial waste recycling or safe disposal already implemented	☹ Risk of metal release in transport ↑ Tightly controlled transport already implemented	☹ Risk of metal release in compounding ↑ Controlled compounding with post-industrial waste recycling or safe disposal already implemented	☺ Risk not perceived as metals are immobile in PVC compound during product	☹ Risk of metal release in incautious disposal ☺ Increasing controlled-loop recycling in Europe means that metals are contained for reuse

(Continued)

TABLE 5.3 (Continued) Potential sustainability issues by TNS System Condition associated with the use of metal-based stabiliser additives to PVC, including potential positive (☺) and negative (☹) implications, and innovation and mitigation measures (↑) at specific points in the final PVC window profile life cycle

	LIFE CYCLE STAGES				
RAW MATERIALS ACQUISITION	ADDITIVE PRODUCTION	PACKAGING AND DISTRIBUTION	COMPOUNDING/ CONVERTING	PRODUCT USE/ MAINTENANCE	POST-USE MANAGEMENT
TNS System Condition 2					
☹ Risks from exhaust fumes and oils during extraction ↑ Use low-emission machinery that is well maintained	☹ Risks from exhausts or other emissions in manufacturing ↑ Low-emission manufacturing with post-industrial waste recycling reducing overall production processing	☹ Emissions in transport ↑ Tightly controlled transport to minimise risk	☹ Risks from emissions during compounding ↑ Use low-emission machinery that is well maintained, with post-industrial waste recycled or safely disposed	☺ The resilience of stabilised PVC gives long service life with low maintenance, averting inputs during extended service life	☹ Risk of emissions in incautious disposal ☺ Controlled-loop recycling means emissions and material value are substantially contained ↑ Further planned progress with controlled-loop recycling will further reduce risks of emissions

(Continued)

TABLE 5.3 (Continued) Potential sustainability issues by TNS System Condition associated with the use of metal-based stabiliser additives to PVC, including potential positive (☺) and negative (☹) implications, and innovation and mitigation measures (↑) at specific points in the final PVC window profile life cycle

	LIFE CYCLE STAGES					
	RAW MATERIALS ACQUISITION	ADDITIVE PRODUCTION	PACKAGING AND DISTRIBUTION	COMPOUNDING/ CONVERTING	PRODUCT USE/ MAINTENANCE	POST-USE MANAGEMENT
TNS System Condition 3	☹ Mining risks habitat loss and groundwater disturbance ↑ Ideally, switch to recycled metal sources, avoiding physical impacts from virgin extraction ↑ For virgin metals, switch to suppliers certified for best mining practice	☹ Habitat displacement by manufacturing operations, and risks to water systems ↑ Seek water- and biodiversity-neutral design of plant and management systems	☹ Physical disruptions from transport ↑ Tightly controlled transport with full loads to minimise overall transport movement and their risks ↑ Select packaging that may be reusable or will not be landfilled	☹ Habitat displacement by compounding operations, and risks to water systems ↑ Seek water- and biodiversity-neutral design of plant and management systems ↑ Ensure that waste is minimised and diverted from landfill	☺ The resilience of stabilised PVC gives long service life with low maintenance inputs (including water use), averting management activities and inputs damaging to nature	☹ Risk of emissions in incautious disposal physically damaging ecosystems ☺ Controlled-loop recycling means material value is contained and landfill is not required ↑ Further planned progress with controlled-loop recycling will reduce risks of physical damage

(Continued)

TABLE 5.3 (Continued) Potential sustainability issues by TNS System Condition associated with the use of metal-based stabiliser additives to PVC, including potential positive (☺) and negative (☹) implications, and innovation and mitigation measures (↑) at specific points in the final PVC window profile life cycle

	LIFE CYCLE STAGES					
	RAW MATERIALS ACQUISITION	ADDITIVE PRODUCTION	PACKAGING AND DISTRIBUTION	COMPOUNDING/ CONVERTING	PRODUCT USE/ MAINTENANCE	POST-USE MANAGEMENT
TNS System Condition 4*	☹ Potential health risks and unfair shares of revenues during extraction ↑ Ensure that suppliers are audited for low-emission machinery and safe and ethical operating practices	☹ Potential health risks during manufacturing ↑ Ensure that manufacturing uses safe and well-maintained machinery with safe and ethical operating practices	☹ Potential health risks during transport ↑ Ensure that transport uses safe and well-maintained machinery with safe and ethical operating practices	☹ Potential health risks during compounding ↑ Ensure that compounding uses safe and well-maintained machinery with safe and ethical operating practices	☺ The resilience of stabilised PVC gives long service life with low maintenance, supporting the meeting of diverse human needs during extended service life	☹ Health risks through incautious disposal ☹ Waste of material value through incautious disposal ☺ Controlled-loop recycling means material value is contained and safety is ensured ☺ Controlled-loop recycling also creates new employment opportunities

'Traffic lights' colour coding of cells indicates positive benefits with issues already implemented (green), a balance of issues to address and benefits or

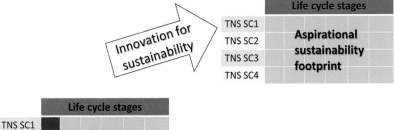

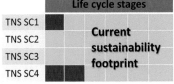

FIGURE 5.9 Illustrative representation of how a 'heat map' of priority areas of unsustainability revealed by ASF assessment can inform areas requiring innovation to achieve the ultimate aim of the sustainable use of additives and other substances. ©Mark Everard.

to PVC and vinyl products but to all other materials in societal use. In this section we will look at a subset of the novel PVC life cycles driven by sustainable development challenges.

Of the many innovations in PVC polymers, BIOVYN™ is worth noting as a major step forward towards carbon neutrality.[19] BIOVYN, as suggested by the name, is a bio-attributed PVC. The novel contribution of BIOVYN, launched in 2019 by the company INEOS Inovyn (part of the INEOS group), is that the PVC polymer is made from 100% renewable organic feedstock. This positions INEOS Inovyn as the world's first commercial producer of bio-attributed PVC. Innovation of BIOVYN was driven by commitments to sustainable development informed by ongoing pursuit of the TNS Sustainability Challenges for PVC.

The renewable supply chain from which BIOVYN is derived is fully certified by the Roundtable on Sustainable Biomaterials,[20] a global multi-stakeholder membership organisation established to drive sustainable transition to a bio-based and circular economy. Significantly, BIOVYN offers customers a reduction in carbon footprint of over 90% relative to traditional petrochemically derived virgin PVC. BIOVYN PVC also makes a significant contribution to the circular economy through making use of secondary biological resources that do not compete with the food chain, and might otherwise be regarded as waste from primary resource extraction. BIOVYN is of equivalent quality, performance and recyclability to PVC polymers manufactured by alternative routes. It is equally amenable to durable, long-life applications, but offers a substantially improved sustainability footprint for products in the progression

TABLE 5.4 Achievements against VinylPlus targets grouped the five TNS Sustainability Challenges for PVC, and their inclusion under the VinylPlus Pathways to 2030

TARGET	REPORTED PROGRESS TO 2022
VinylPlus sustainability challenge 1. Controlled-loop management (now under VinylPlus Pathway 1: 'Circular economy')	
"VinylPlus is committed to recycling at least 900,000 tonnes of PVC waste into new products by 2025 and 1 million tonnes by 2030".	The *VinylPlus Progress Report 2023*[15] reports that 813,266 tonnes of PVC waste were recycled in 2022, aiming for 900,000 tonnes by 2025 and 1 million tonnes by 2030. A total of 8.1 million tonnes of PVC were reported as being recycled since 2000, with an associated 16.2 million tonnes of CO_2 saved and the creation of 1,600 direct jobs in recycling plants. The amount of PVC waste recycled represented around 27% of the total PVC waste generated in 2022 in the EU-27, Norway, Switzerland and the UK. Demand for recycled PVC (rPVC) remained very high, the registered uptake of rPVC from converters reaching 561,795 tonnes in 2022: a 24.8% increase compared to 2021 despite a downturn in demand in late 2022, which was attributed to spiralling energy costs. (Reuse of by-products is not considered recycling, and it is not counted towards the recycling targets.) The BIOVYN product,[16] launched in 2019 based on 100% renewable organic feedstock, marks significant progress with over 90% saving on carbon footprint relative to virgin PVC produced from petrochemical sources.
VinylPlus sustainability challenge 2. Organochlorine emissions (now under VinylPlus Pathway 2: 'Decarbonisation and environmental footprint minimisation')	
The European Council of Vinyl Manufacturers 'ECVM Charter', first established in 1995 and subsequently updated, drive continuous improvement at the manufacturing phase.	The *VinylPlus Progress Report 2023* reports on the requirement for plants operated by European PVC manufacturers to comply with high standards of control over processes, operations and emissions under the 'ECVM Charter',[17] subject to third-party certification by DEKRA GmbH,[18] with an overall compliance rate of 89% in 2022. Increasing circularity driven by actions under TNS Sustainable Principles 1 (Controlled loop) and 5 (Sustainability awareness) continue to address potential emissions of organochlorine and other persistent substances throughout the life cycles of PVC products.

(Continued)

TABLE 5.4 (Continued) Achievements against VinylPlus targets grouped the five TNS Sustainability Challenges for PVC, and their inclusion under the VinylPlus Pathways to 2030

TARGET	REPORTED PROGRESS TO 2022
VinylPlus sustainability challenge 3. Sustainable use of additives (now under VinylPlus Pathway 2: 'Decarbonisation and environmental footprint minimisation')	
"The Additive Sustainability Footprint® (ASF) is a methodology to proactively assess and promote the sustainable production and use of PVC additives throughout entire product lifecycles, including the roles of additives in the performance of PVC products".	The *VinylPlus Progress Report 2023* reports achievement of the targets: 1. "By 2021, organisation of at least one introductory ASF webinar by VinylPlus" (target met); 2. "By 2022, produce a report on the sectors' / partners' experience and application of the ASF tool" (target met). Reports have been produced on ASF assessments of two EU-generic products (a window profile; and homogeneous PVC flooring) by end-2022. In addition, the VinylPlus Supplier Certificate (VSC) has been produced, and five additive (stabiliser) manufacturing companies – Reagens; Baerlocher; Akdeniz Chemson; IKA; and Polymer Chemie – have become VSC-accredited by end-2022. Additional additive manufacturers are going through the VSC auditing process at the time of writing. Additional commitments have been made in 2023 under the VinylPlus Sustainable Additives Committee to monitor and address international, European and national regulatory changes that may lead to restriction or authorisation for legacy PVC additives in waste streams. These additional VinylPlus Sustainable Additives Committee commitments also address other emergent issues, which include: undertaking further human health and environmental risk assessment studies; socio-economic studies to address knowledge gaps justifying exemptions on restrictions or authorisations; and developing evidence-based arguments for a sustainable migration path resolving conflicts between EU 'clean chemistry' and 'circular economy' strategies. There is also interest in further developing digital tools making ASF assessment easier, quicker and cheaper; testing the ASF methodology in different PVC applications through generic assessments; and undertaking training and support for VinylPlus companies to run company-level ASF assessments as well as promoting ASF outside Europe.

(Continued)

TABLE 5.4 (Continued) Achievements against VinylPlus targets grouped the five TNS Sustainability Challenges for PVC, and their inclusion under the VinylPlus Pathways to 2030

TARGET	REPORTED PROGRESS TO 2022
VinylPlus sustainability challenge 4. Sustainable use of energy and raw materials (now under VinylPlus Pathway 2: 'Decarbonisation and environmental footprint minimisation')	
The *VinylPlus Progress Report 2023* states that: 1. "VinylPlus will evaluate the potential and, by 2025, report on projected core carbon reduction progress to be achieved by 2030". 2. "By 2025, report on the use of renewable energy". 3. "By 2025, report on sustainable feedstock sourcing".	The *VinylPlus Progress Report 2023* reports a 9.5% reduction in energy consumption per tonne of PVC in 2015-2016 compared to 2007-2008, with a 14.4% reduction in CO_2 emissions, and between 16% and 26.5% reductions in energy consumption for window profiles, pipes, flooring, films and sheets in 2020 compared to 2010. All members of the European Stabiliser Producers' Association (ESPA) Council are reviewing how to adopt a carbon footprint analysis for their additive products. Across VinylPlus, there is a focus on developing a carbon handprint (measures delivering positive progress) approach, building on the carbon footprint (total emissions) assessments. Carbon emissions feature at product level (under the VinylPlus Product Label) and additive substance level (under the Additive Sustainability Footprint and the VinylPlus Supplier Certification schemes). The >90% saving on carbon footprint reported for the BIOVYN product relative to virgin PVC produced from petrochemical sources (see under VinylPlus sustainability challenge 1 above) also makes a very substantial contribution to this energy/decarbonisation challenge. *(Continued)*

TABLE 5.4 (Continued)   Achievements against VinylPlus targets grouped the five TNS Sustainability Challenges for PVC, and their inclusion under the VinylPlus Pathways to 2030

TARGET	REPORTED PROGRESS TO 2022
VinylPlus sustainability challenge 5. Sustainability awareness (now under VinylPlus Pathway 3: 'Coalitions and partnerships')	
The *VinylPlus Progress Report 2023* records a range of specific commitments under the broad target headings of: • "Ensuring transparency and accountability; • "Contributing to sustainable development through certified and traceable products; • "Engaging stakeholders in the sustainable transformation of the PVC industry; and • "Partnering with stakeholders".	The *VinylPlus Progress Report 2023* lists a wide range of initiatives at European, national and local levels to raise awareness of PVC sustainability and the VinylPlus 2030 Commitment. As one example, VinylPlus Partners have engaged with institutions, authorities, local administrations, and Olympic organisers involved in the 2026 Winter Olympics Milano-Cortina triennial project in Italy since 2020 "...to raise awareness of how PVC can contribute to a sustainable sporting event, thanks to its sustainability and recyclability features, as well as its technical and economic characteristics". VinylPlus continues to publish annual, independently audited VinylPlus Progress Reports, noting contributions from each VinylPlus industry sector. As for other sustainability challenges, further specific details can be found in the *VinylPlus Progress Report 2023*, and future annual updates. These include, for example: progress with roll-out of the VinylPlus® Product Label and its acceptance by green building standards and procurement systems; wider attainment of the VinylPlus Supplier Certificate (VSC); reporting on how PVC products contribute to climate change reduction; engagement with international and intergovernmental organisations to share VinylPlus' knowledge, experience and business model for sustainability; and contributions to projects and partnerships supporting attainment of the UN Sustainable Development Goals (SDGs).

towards a circular, carbon-neutral economy. Since its launch in 2019, there has been global interest in BIOVYN across most industrial sectors as pursuit of sustainable development goals, and particularly the decarbonisation agenda, has risen in political and business priorities.

The PVC polymer is, as we have seen, compounded with a wider range of additives imbuing the finished PVC plastic compound with a wide diversity of desirable properties. In fact, the PVC polymer may often be less than 50% by mass of the finished plastic used for some applications, flooring in particular including a substantial amount of inert filler. Innovations in PVC additives and their use therefore play major roles in making progress towards overall sustainability.

The voluntary phase-out of problematic heavy metals in PVC stabilisers in Europe, particularly cadmium and lead, is one example of the industry recognising collectively that these elements are not part of a vision of a sustainable future and, informed by backcasting, taking elective action to replace them. The same is also true of some inherently problematic plasticiser substances, particularly short-chain phthalates such as Di (2-ethylhexyl) phthalate (DEHP), which not only have some known contributions to birth defects and infertility in animals but can also leach from PVC into fluids such as blood or nutrition formulas. Consequently, plastic items containing more than 1% DEHP are banned in plastic items designed for children – such as toys, plastic figures, some sporting good, childcare articles, eating vessels and utensils that are readily chewed and/or sucked – in many territories across the world.[21] DEHP is also now restricted more generally under EU REACH regulations. Nevertheless, DEHP is still permitted in the UK in some medical devices (such as intravenous (IV) tubing, umbilical artery catheters, and blood bags and infusion tubing) where flexibility makes them easier to use, less likely to damage body tissues and more comfortable for patients. The UK's Medicines & Healthcare Product Regulatory Agency recognises that there is inconclusive or inconsistent evidence to suggest that medical devices containing DEHP pose an unacceptable health risk to humans (specifically male reproductive health) and therefore, balanced with the fact that medical devices containing PVC plasticised with DEHP have important clinical benefits, has not imposed a ban on this application but recognises a need to improve on current knowledge about the toxicological profile of DEHP and also to develop as soon as possible alternative materials with a favourable profile for both efficiency and safety.[22] Nonetheless, further work is still required with respect to some additive substances. For example, a 2023 study conducted on behalf of the Norwegian Environment Agency in which a range of plastic consumer products – not limited to PVC – were cut into smaller pieces and immersed in a mixed acetone/ hexane solvent under ultrasonication, with a follow-up study taking surface wipe samples, explored leaching of additives.[23] Some additives (chlorinated paraffins and brominated flame retardants) were found to have been released in detectable concentrations in some applications from a range of plastic types under these quite extreme testing regimes, nonetheless indicating potential for health impacts.

Innovation of additives by many supplier companies not only includes the replacement of formerly problematic substances, for example replacement of lead-, cadmium- and now tin-based stabilisers with calcium/zinc and calcium/organic systems, but also with a view to retention of stabiliser functionality when the plastic compound is recycled at the end-of-life of PVC products. Various companies are also innovating additive products for integration into recovered PVC during the recycling phase in order to boost stabilisation and other technical properties, including as 'one-pack' formulations combining multiple additive types.

Innovations to promote the recyclability of PVC compounds at product end-of-life span a range of additive systems, simultaneously addressing linked TNS Sustainability Challenges for PVC including Challenge 1 (Controlled loop), Challenge 3 (Sustainable use of additives), Challenge 4 (Sustainable energy and climate stability, due to the substantial carbon benefits of recycling) and Challenge 5 (Sustainability awareness, by engaging more players in the PVC value chain). Logistical innovations by some additive manufacturers can also yield sustainability benefits at other points along the value chain, such as the Spanish additive manufacturer ASUA giving discounts for customers buying full loads of additive products and thereby reducing carbon emissions and packaging during the transport phase, as well as agreements to take back unused products from customers for beneficial reuse without waste generation.

Pursuit of circularity has many linked benefits not just for delivery of Challenge 1 (Controlled loop) but across all five of the TNS Sustainability Challenges for PVC. The *VinylPlus Progress Report 2023*[24] emphasises efforts by players along the entire European value chain to advance circularity, enabling the recycling of nearly 8.1 million tonnes of PVC into new products between 2000 and 2022 and preventing the release of more than 16.2 million tonnes of CO_2 into the atmosphere. Significant initiatives include Recovinyl®, established in 2003 by VinylPlus as part of the Vinyl2010 Voluntary Commitment and now spanning 150 recycler partners. Recovinyl monitors and verifies the recycling of PVC waste and uptake of recycled PVC into new products, developing recycling technology, boosting the collection and recycling of PVC waste, and promoting demand for recycled PVC material from the converting industry. Within Recovinyl, RecoTrace™ has been developed to further enhance recording and tracing schemes for recycling volumes and the uptake of recyclates into new products. RecoTrace is also the first system to comply with the monitoring requirements of the Circular Plastics Alliance,[25] established in 2019 to support the growing demand of major players in diverse industries to create a closed loop in their logistic chain. Across the UK, the Recofloor™ programme, founded by flooring manufacturers Altro and Polyflor, operates as a collection and recycling service for commercial waste vinyl flooring as an eco-friendly and cost-effective alternative to traditional waste disposal routes, helping the flooring industry and its partners progress towards a more sustainable, 'net zero' cyclic approach.[26] Members of the European Resilient Flooring Manufacturers' Institute (ERFMI) recycled more than 130,000 tonnes of PVC in 2019 verified by RecoTrace.[27] The VinylPlus Product Label, the Additive Sustainability Footprint (ASF) and the VinylPlus Supplier Certificate (VSC) schemes also include criteria reflecting the sustainable benefits of circularity.

Research continues in Europe to drive greater circularity. Just two examples are the Circular Flooring Project and REMADYL. Neither research project is yet complete at the time of writing, but both serve as examples of ongoing investment in addressing more difficult aspects of making practical progress with the sustainability challenges.

The Circular Flooring Project, funded by the EU, aims to establish a circular recycling process for plasticised PVC from post-consumer waste flooring, extracting and converting 'legacy' additives into REACH-compliant products and thereby diverting flooring materials from landfill or incineration and avoiding the loss of valuable resources.[28] It thereby reduces CO_2 emissions and other environmental burdens associated with waste disposal by establishing closed recycling loops for various material streams. One innovation under the Circular Flooring Project is development of the CreaSolv® recycling process to separate the PVC resin from post-consumer floor coverings containing legacy plasticisers (phthalic acid esters) that are not REACH-compliant, enabling the recycling of the PVC content into REACH-compliant alternatives including reuse in new floor coverings as a contribution to a circular European economy. The Circular Flooring Project received funding from the European Union's Horizon 2020 research and innovation programme.

The REMADYL research project also aims to develop a process to recycle PVC from post-consumer rigid and plasticised products containing legacy additives that are now non-compliant with REACH, particularly low molecular weight phthalate plasticisers (mainly DEHP) and heavy metal-based stabilisers (mainly lead).[29] REMADYL seeks to develop a breakthrough continuous process based on extractive extrusion technology in combination with novel solvents and melt filtration, aimed at 'rejuvenating' the polymer into a high-purity PVC compound usable in soft and hard PVC products at competitive cost.

Whilst the various certification schemes, standards, challenges and pathways created by VinylPlus are important to inform and monitor practical progress, it is important to remember – as indeed I remind the industry on a regular basis – that they are a means to an end of driving progress with sustainable development by backcasting from principles of sustainability, as framed by The Natural Step approach. And, with new entrant companies and staff turnover, it is vital from time to time to refresh the narrative of how fundamental scientific principles (such as the principle of matter conservation, the laws of thermodynamics and aspects of cell metabolism) underpin a non-contentious model of the workings of the planetary biosphere, from which TNS System Conditions are derived informing challenges for PVC value chains, and that these in turn are integrated into VinylPlus Pathways and the various certification schemes. These certification schemes are far from simplistic static standards; they are tools informed by the strategic purpose of propelling innovations and progress towards the goal of sustainability.

5.9 ELSEWHERE IN THE WORLD

The PVC industry in other regions of the world has grappled with sustainable development challenges in different ways. A very brief overview of international commitments and progress includes:

- **North America**. The Vinyl Sustainability Council was established in North America. Part of its mission is recognising the business importance of sustainability in meeting the needs of a growing global population faced with depleting natural resources.[30] The Vinyl Sustainability Council was created to advance the efforts of the vinyl value chain across North America under which members collaborate to develop and implement best practices and innovation, and to promote achievements to key stakeholders. The Council has also created a Vantage Vinyl brand[31] to indicate that vinyl products have been verified, recognising an achievement in sustainability.
- **India**. An Indian Vinyl Council[32] has been established with a range of objectives including development of the whole vinyl value chain and its positive public image, with work strands engaging with government and non-government bodies and statutory authorities in the formulation of policies, codes and standards. There is a focus on promotion and support of standardisation and quality assurance programmes to encourage regulatory compliance, and promotion with end-users of the "...value proposition of PVC products including energy conservation, eco-friendliness and sustainability". Additional objectives relate to fostering innovation, training and skills development within the value chain and the raising of

the PVC industry to global standards, promotion of scientific and economic research and broadening the market for PVC products. In addition to these stated objectives, deeper dialogue is occurring about commitments to sustainable development of PVC and the PVC sector, some of which entails inputs from players connected with VinylPlus.

- **ASEAN**. The ASEAN Vinyl Council (AVC), operating across the ASEAN (Association of Southeast Asian Nations) region, states that "…members commit that its environmental impact shall be precisely measured, monitored, and reduced to meet reference targets and timeframes which will be determined periodically by the AVC Steering Committee".[33]
- **Australia**. The Vinyl Council of Australia[34] represents the PVC/vinyl value chain in Australia, working to advance the sustainability of PVC products and the industry by encouraging members to implement product stewardship and best practice manufacturing. The Vinyl Council of Australia also recommends PVC products from companies signed up to its PVC Stewardship programme or accredited as Best Practice PVC.
- **Southern Africa**. The Southern African Vinyls Association[35] (SAVA) is the voluntary industry association representing the entire value chain in the sub-Saharan region, helping its members to create and grow a dynamic, sustainable and responsible vinyls industry, and communicating the latest, scientifically-based and researched information.
- **Additional national PVC institutions**. National institutions not noted above include: The Instituto Brasileiro do PVC (https://pvc.org.br/); the Vinyl Institute of Canada (https://www.vinylinstituteof-canada.com/); and the Asociacion Argentina del PVC (https://www.aapvc.org.ar/).
- **International**. The Global Vinyl Council (GVC)[36] was established to promote the value of PVC/vinyl products and represent its members in advocacy actions world-wide. The GVC has a range of aims, many related to aspects of sustainable development. Relevant stated aims include the following:

  - Identify and promote technologies and practices that will further improve the health, safety and environmental performance of the vinyl industry.
  - Support performance standards, 'best practice' models and successful programs that ensure all production facilities around the world operate to the highest standards.

- Fund and manage world-wide studies and programs supporting its objectives.
- Foster cooperation between member companies, governments and organizations to increase the understanding of vinyl and the industry.
- Advocate responsible lifecycle management (in the manufacturing, use, recycling and disposal) of vinyl products.
- Promote the value of vinyl products to society. Communicate the benefits of vinyl products, and address related issues and concerns.
- Avoid discrimination and de-selection of vinyl products and ensure that a level playing field is maintained.
- Provide technical information to support its members.
- Provide its members with a forum where they can seek information and support, and share experiences.

The International Council of Chemical Associations (ICCA: www.icca-chem. org), the trade association of the global chemical industry representing chemical companies accounting for more than 75% of global production capacities, is another significant body with interests, amongst its wide remit, of progressing sustainable development of the plastics sector. ICCA is the steward of the Responsible Care programme, but also has a wide range of working groups. One of these is the ICCA Plastics Additives Working Group, recognising the industry's willingness to be more transparent and address external concerns, particularly concerning health and safety, and to make progress towards the elimination of plastic waste. This includes recognition of a need to develop tools that are accepted and supported by authorities and other stakeholders informing improved management.

It is part of the mission of VinylPlus under Pathway 3 to internationalise the approaches and successes achieved in Europe, recognising that the PVC sector is not only a global industry but that supply and value chains also span continents. It is also increasingly evident that the pressing challenges facing the world are fully global in scale, and cannot be met with regional activities alone however successful they may be within the home range. As noted earlier in this book, Europe leads the way with recycling of all types of plastics at 30% of production volumes, albeit leaving substantial room for improvement, but at a global scale there is only an estimated 9% recycling rate including close to zero across much of the developing world. PVC recycling in Europe was in advance of this general plastics rate in 2022, recycling 813,266 tonnes of PVC waste of which 62.42% was pre-consumer waste and 37.58% post-consumer waste.[37] The production of PVC stabilisers containing lead was fully discontinued by voluntary commitment in Europe in 2015, yet lead stabilisation still occurs in several other parts of the world, such as in Asia and in Africa, lead-ing to persistence of this harmful metal in PVC products produced outside Europe also with the possibility of them being imported. Less well-managed PVC manufacturing processes still operate in some global regions, potentially harming health and affecting the reputation of the industry even in regions where emissions are tightly controlled.

On invitation from overseas bodies, I have given a wide range of key-note conference addresses and other corporate and trade association presenta-tions across European countries. So too in India where I have been invited to give keynote addresses, as well as opportunistically making presentations when visiting on my other natural resource management research and advi-sory visits. The same is true in the ASEAN region, mixing invited keynotes with opportunist presentations whilst in country. I have also had meetings with senior people in the South African PVC industry when working in country as a water management advisor. In the US, I have given a number of keynote talks. The same would have been true in Australia, though regrettably Covid-19 thwarted my invitations in 2020 and again in 2021.

A particular concern that I have is that products bought online that are made in places with little scrutiny of chemistry, pollution and ethics in supply chains and manufacturing evade regulatory control when crossing into Europe. Researcher colleagues of mine in the UK have shown me worrying analyses of a number of product types bought through popular trading websites. This is a loophole that needs to be plugged if responsible PVC production and use is to become mainstream, and also to ensure that positive, pro-sustainability efforts in Europe are not undermined.

NOTES

1 ECVM. (n.d.). ECVM's Charter: ECVM's Plants Conform to High Standards of Control over Processes, Operations and Any Emissions to the Natural Environment. European Council of Vinyl Manufacturers (ECVM). [Online.] https://pvc.org/sustainability/industry-responsible-care/ecvm-charter/, accessed 18 March 2023.
2 United Nations. (2015). *The 17 Goals*. United Nations. [Online.] https://sdgs.un.org/goals, accessed 13 March 2023.
3 VinylPlus. (2015). *Our Contribution to Sustainability*. VinylPlus. [Online.] https://www.vinylplus.eu/sustainability/our-contribution-to-sustainability/, accessed 13 March 2023.
4 Everard, M. and Longhurst, J.W.S. (2018). Reasserting the primacy of human needs to reclaim the 'lost half' of sustainable development. *Science of the Total Environment*, 621, pp. 1243–1254. DOI: https://doi.org/10.1016/j.scitotenv.2017.10.104.

5 World Commission on Environment and Development. (1987). *Our Common Future*. Oxford University Press.

6 Everard, M. (2017). Repurposing business around the meeting of human needs. *Environmental Scientist*, pp. 40–45.

7 VinylPlus. (n.d.). *VinylPlus® Product Label*. VinylPlus. [Online.] https://productlabel.vinylplus.eu/wp-content/uploads/2021/07/Product-Label-Brochure-EN_-8JUNE2021.pdf, accessed 13 March 2023.

8 PEFC. (n.d.). *Programme for the Endorsement of Forest Certification*. PEFC. [Online.] https://www.pefc.co.uk/, accessed 13 May 2023.

9 Repubblica Italiana. (2022). Criteri Ambientali Minimi per L'affidamento del Servizio di Progettazione Ed Esecuzione Dei Lavori Di Interventi Edilizi. *Gazzetta Ufficiale Della Repubblica Italiana*, Anno 163° - Numero 183. [Online.] https://www.gazzettaufficiale.it/eli/gu/2022/08/06/183/sg/pdf, accessed 29 April 2023.

10 VinylPlus. (n.d.). *VinylPlus® Supplier Certificate: Criteria for PVC Additives Suppliers*. VinylPlus. [Online.] https://productlabel.vinylplus.eu/wp-content/uploads/2022/07/VinylPlus-Supplier-Certificate_Criteria-Scheme-additives-suppliers-20211105-002.pdf, accessed 13 March 2023.

11 Lundholm, K., Blume, R. and Oldmark, J. (2011). *Process Guide to Sustainability Life Cycle Assessment - A strategic approach to assessing the life cycle of product systems using the Framework for Strategic Sustainable Development*. The Natural Step International, Stockholm.

12 ISO. (2006). Environmental Management – Life Cycle Assessment. International Organization for Standardization, Geneva. [Online.] https://www.iso.org/standard/37456.html, accessed 13 March 2023.

13 Everard, M. and Blume, R. (2019). Additive sustainability footprint (ASF): rationale and pilot evaluation of a tool for assessing the sustainable use of PVC additives. *Journal of Vinyl and Additive Technology*, 26(2), pp. 196–208. DOI: https://onlinelibrary.wiley.com/doi/10.1002/vnl.21733.

14 Everard, M. (2022). Assessment of the sustainable use of chemicals on a level playing field. *Integrated Environmental Assessment and Management*, 19(4), pp. 1131–1146. DOI: https://doi.org/10.1002/ieam.4723.

15 VinylPlus. (2023). *Progress Report 2023: Reporting on 2022 Activities*. VinylPlus. [Online.] https://www.vinylplus.eu/wp-content/uploads/2023/05/VinylPlus_ProgressReport_23.pdf, accessed 14 May 2023.

16 INEOS Inovyn. (2023). *BIOVYN – INOVYN Bio-attributed PVC: A Big Step Forward Towards Carbon Neutrality*. INEOS Inovyn. [Online.] https://biovyn.co.uk/, accessed 14 May 2023.

17 ECVM. (2023). *ECVM's Charter*. European Council of Vinyl Manufacturers (ECVM). [Online.] https://pvc.org/sustainability/industry-responsible-care/ecvm-charter/, accessed 31 March 2023.

18 Dekra. (2023). *DEKRA at a Glance*. Dekra. [Online.] www.dekra.com, accessed 14 May 2023.

19 INEOS Inovyn. (2023). *BIOVYN – INOVYN Bio-attributed PVC: A Big Step Forward towards Carbon Neutrality*. INEOS Inovyn. [Online.] https://biovyn.co.uk/, accessed 14 May 2023.

20 RSB. (2023). *What Is the RSB?* Roundtable on Sustainable Biomaterials (RSB). [Online.] https://rsb.org/, accessed 13 May 2023.

21   For example: Commonwealth of Australia. (2010). *Permanent Ban on Children's Products Containing More That 1% Diethylhexyl Phthalate (DEHP)*. Commonwealth of Australia, Competition and Consumer Act 2010, Consumer Protection Notice No. 11 of 2011. [Online.] https://www.legislation.gov.au/Details/F2011L00192, accessed 01 April 2023.

22   Medicines & Healthcare Product Regulatory Agency. (2021). *Guidance: DEHP Phthalates in Medical Devices, Updated 26 January 2021*. Medicines & Healthcare Product Regulatory Agency. [Online.] https://www.gov.uk/government/publications/dehp-phthalates-in-medical-devices/dehp-phthalates-in-medical-devices, accessed 01 April 2023.

23   Bohlin-Nizzetto, P. (2023). *Content and Migration of Chemical Additives from Indoor Consumer Plastic Products*. Norwegian Institute for Air Research (NILU), NILU report 6/2023.

24   VinylPlus. (2023). *Progress Report 2023: Reporting on 2022 Activities*. VinylPlus. [Online.] https://www.vinylplus.eu/wp-content/uploads/2023/05/VinylPlus_ProgressReport_23.pdf, accessed 14 May 2023.

25   Circular Plastics Alliance. (2023). *About Us*. Circular Plastics Alliance. [Online.] https://www.circular-plastics-alliance.com/en/about-us/, accessed 14 May 2023.

26   Recofloor. (n.d.). *Recofloor Vinyl Take-Back Scheme*. Recofloor.org. [Online.] https://www.recofloor.org/, accessed 29 April 2023.

27   ERFMI. (2023). *About. Our Purpose*. European Resilient Flooring Manufacturers' Institute (ERFMI). [Online.] https://erfmi.com/#environment, accessed 13 May 2023.

28   Circular Flooring Project. (2022). *New Products from Waste PVC Flooring and Safe End-of-Life Treatment of Plasticisers*. Circular Flooring Project. [Online.] https://www.circular-flooring.eu/circular-flooring-project/, accessed 14 April 2023.

29   REMADYL. (n.d.). *Objectives*. REMADYL. [Online.] https://www.remadyl.eu/project/objectives/, accessed 14 April 2023.

30   Vinyl Institute. (n.d.). *Vinyl Sustainability Council*. Vinyl Institute. [Online.] https://www.vinylinfo.org/vinyl-sustainability-council/, accessed 15 March 2023.

31   Vantage Vinyl. (n.d.). *Vantage Vinyl for a Sustainable Future*. Vantage Vinyl. [Online.] https://vantagevinyl.com/, accessed 15 March 2023.

32   Indian Vinyl Institute. (n.d.). *Objectives*. Indian Vinyl Institute. [Online.] http://indianvinylcouncil.com/, accessed 15 March 2023.

33   ASEAN Vinyl Council. (n.d.). *Our Commitment*. ASEAN Vinyl Council. [Online.] http://aseanvinyl.com/sustainability/our-commitment/, accessed 15 March 2023.

34   Vinyl Council of Australia. (n.d.). *The Vinyl Council of Australia Represents the PVC/Vinyl Value Chain in Australia*. Vinyl Council of Australia. [Online.] https://vinyl.org.au/, accessed 15 March 2023.

35   Southern African Vinyls Association. (2023). *The Southern African Vinyls Association (SAVA)*. Southern African Vinyls Association (SAVA). [Online.] https://savinyls.co.za, accessed 01 April 2023.

36   Global Vinyl Council. (n.d.). *The Global Vinyl Council (GVC)*. Global Vinyl Council. [Online.] https://www.vinylinfo.org/the-global-vinyl-council/, accessed 15 March 2023.

37   VinylPlus. (2023). *Progress Report 2023: Reporting on 2022 Activities*. VinylPlus. [Online.] https://www.vinylplus.eu/wp-content/uploads/2023/05/VinylPlus_ProgressReport_23.pdf, accessed 14 May 2023.

A Level Playing Field?

6

Within the title of this book, and repeated throughout, is the phrase 'level playing field'. In a world of increasing human demands on dwindling resources, there is a pressing need for a scientifically founded basis for determining which materials can best serve our needs in the most efficient and safest manner accounting for the full life cycles of the products we use. This level of objectivity is not something that PVC has enjoyed to any great extent as the heightened attentions of many environmental NGOs, media coverage and some actions and instruments of regulatory bodies have been disproportionate relative to other bulk materials in common societal use.

The lack of equivalent comparison of the sustainability virtues of *Alternatives to PVC*[1] suggested by Greenpeace, as also a suspicion in some cases of collusion by manufacturers of competitor materials, is highlighted earlier in this book. So too are decisions by some organisations to deselect PVC in the absence of any critical appraisal of substitute materials that are, implicitly and largely wrongly, automatically and naively assumed to be 'good'. The fact that these reactive yet uncritical substitution decisions are falsely claimed to be part of a commitment to sustainable development is entirely disingenuous. Such decisions are also ultimately risky for businesses that may simply be reinvesting in tomorrow's as-yet unresearched problems. Surely any business or institution would want to underpin its claimed commitment to sustainable development by ensuring that the decisions it has made genuinely represent stepwise increments towards clear end-goals, rather than reactively sinking investment blindly into untested materials and/or processes that might incur unforeseen liabilities and conceivably a new wave of NGO campaigning in future?

Consequently, the clarion cry for a level playing field of assessment rings throughout this book, not as a means to support the case for PVC, or indeed any other material, but as a scientifically grounded basis for intelligent innovation and decision-making about which of the huge variety of substances used by society can best meet specific human needs in the safest and most efficient manner. It is also a springboard for innovations and improvements

DOI: 10.1201/9781003453949-6

across material life cycles representing sustainable improvements, informed by assessments against that level playing field.

Contention remains in various reports and news items about PVC. The European Commission's 2000 Green Paper *Environmental Issues of PVC*[2] highlighted several issues of concern, principally associated with certain additives and the need to increase recycling of PVC. In the US, the EPA has paid significant attention to emissions of substances, including VCM, during the manufacture of PVC, as well as identifying concerns about some additives. Conversely, the 1997 NCBE study[3] found no compelling scientific reason for retailers not to continue to use PVC whilst the 2004 European Commission study *Life Cycle Assessment of PVC and of principal competing materials*[4] found that the environmental profiles of PVC products were broadly similar than those for other substances in equivalent durable applications, both cited and explained in Chapter 4, 'PVC: The Good, the Bad and the Prejudiced' including their recognition that careful manufacture, use, recycling and final disposal of PVC products to the highest standards was required to control risks.

The full attainment of sustainability remains a perhaps distant goal in relation to the life cycles of the diverse range of materials used by society to service its needs today, as for many of society's wider activities and challenges. The TNS Sustainability Challenges for PVC represent a signpost on a far from complete journey towards the goal of sustainability for the PVC sector, including all stakeholders involved in the value chains of PVC products. Significant and tangible progress has been made across the PVC value chain over the decades that have intervened since various problems have manifested, driven in Europe by voluntary commitments based on clearly articulated scientific sustainability challenges, as well as regulatory pressures, with substantial practical actions and investments. PVC, aided by its durable and inherently mechanically recyclable properties, has more recently been recognised as a pioneer of practical progress towards a 'circular economy',[5] particularly in healthcare[6] and for durable building and construction products where recovery and recycling systems are now well established. Parallel systemic sustainability assessments are often lacking for other bulk materials used in equivalent applications.

PVC then is not a material apart from all others, as might be inferred by sporadic NGO and media reporting (in some cases with support of unfounded claims by competitor material producers) that still often reference anachronistic issues. This is not to say that PVC does not have continuing challenges, including some specific to its chemistry. But then the same is true for all other materials used in society, which have both their own specific issues but also many generic challenges stemming from the ways in which society still generally lacks coherent management of whole value chains from raw material sourcing through manufacture, use and maintenance, and handling of products at and beyond end-of-life. Nonetheless, PVC has continued to receive an asymmetrically large amount of scrutiny relative to other bulk material used by society, many of which escape equivalent scrutiny with an implicit assumption that they are automatically more benign.

The case for a genuinely level playing field of scrutiny, perception and regulation is pressing if society is to make tangible progress towards oft-stated commitments to sustainable development, which necessarily must include selecting or innovating, on an objective basis, the best, safest and most efficient materials to address human needs.

6.1 WIDER COMPARATIVE ASSESSMENT USING THE FIVE TNS SUSTAINABILITY CHALLENGES FOR PVC

Celebrating progress over the two decades since publication of the five TNS Sustainability Challenges for PVC, I published a paper in the peer-reviewed literature in 2020 titled *Twenty years of the Polyvinyl Chloride (PVC) sustainability challenges*.[7] This paper reported how intense campaigning pressure on the UK PVC industry had been beneficial in highlighting some legitimate problems, forcing the industry into recognition of sustainable development as a defining agenda. The paper also summarised the process of engagement by the UK PVC industry with The Natural Step, leading through to publication of the five TNS Sustainability Challenges for PVC in 2000. Initial engagement with the five TNS Sustainability Challenges for PVC in the UK led through to their progressive uptake across European PVC value chains. They were subsequently adapted as the five sustainability challenges underpinning the VinylPlus voluntary commitment programme. VinylPlus has continued to publish and audit progress against targets associated with the five TNS Sustainability Challenges. Engagement by the entire European PVC value chain, initially joining together under Vinyl2010 and then the explicit adoption of the five challenges under VinylPlus, is a rare but effective example of a successful voluntary commitment by a whole industry value chain.

It was in my 2020 paper that I acknowledged for the first time that, in drafting those five TNS Sustainability Challenges for PVC, I was conscious that simply making them wholly specific to PVC might in the long run only reinforce a perception of PVC as a material apart. The wording of the five TNS Sustainability Challenges was therefore consciously framed in terms generically relevant to all other material types. Whilst, ideally, all materials could beneficially be assessed using the full suite of TNS tools – the Framework for Strategic Sustainable Development (FSSD) – the five TNS Sustainability Challenges for PVC represent a simplified shorthand for rapid material screening on the basis of the underpinning sustainability principles. The 2020 paper consequently contained brief assessments of the sustainability performance of PVC as well as three comparator materials commonly presented as alternatives to PVC, judged on the common basis of the five TNS Sustainability Challenges.

This was not an assessment of overall achievement of sustainability, but of whether: material progress was made, mixed progress was evident or no substantive progress was evident. Findings of this comparative assessment in the 2020 paper are briefly summarised below:

PVC in European production under VinylPlus commitments is used in many applications including durable construction products such as window profiles, pipes and cables.

- The 2020 paper recorded that material progress had been made with PVC under all five TNS/VinylPlus Sustainability Challenges. VinylPlus reporting on audited progress with these challenges is described in previous sections of this book.

Timber/forest-based products commonly used in the European window profile market are recommended for this purpose by Greenpeace as a substitute for PVC.[8] A WWF report[9] rightly concluded of PVC in relation to alternative materials in window profiles that "Opinions on its environmental impact and safety are polarised between the chemicals industry and environmental organisations" with selective use of evidence skewing findings, for example taking full account of the longevity, maintenance inputs and service life of different materials and their potential recyclability/value recovery at the end-of-life of window profiles.

- Whilst some forest products, such as short-life food packaging, can be composted if collected, this is not the case for more durable applications such as window profiles, which tend to be degraded and impregnated with preservatives during manufacture and/or throughout their service life. These additives, in the form of preservative paints, pressure treatments and other coatings integrated into timber during product manufacture or in product life, are generally overlooked in terms of their potential sustainability implications (TNS Challenge 4/VinylPlus Challenge 3), limiting the potential for recovery and recycling for beneficial reuse after end-of-life (TNS Challenge 2/VinylPlus Challenge 1). Whilst forests naturally recycle carbon, timber harvesting and processing entail carbon inputs so this is not a clear benefit, but lack of beneficial recoverability at end-of-life means that significant potential carbon neutrality benefits are lost (TNS Challenge 1/VinylPlus Challenge 4). Whilst timber production is generally cleaner than heavy chemical manufacture, regular additions of preservatives are required in wood profiles to prolong service life, though the Minamata Convention[10] and Gothenburg Protocol[11] are examples amongst a range of regulatory and consensual agreements limiting the use of specified hazardous chemicals in preservatives, making limited progress with TNS Challenge 3/VinylPlus Challenge 2. Significant progress with sustainable forest product production has been made under the Forest Stewardship Council (FSC) scheme, described previously, creating substantial and global differentiated markets for timber and other forest products spanning whole value chains from production to consumer products. However, a lack of join-up across the whole societal value chain is evident when forest-based window profile products reach end-of-life, most commonly in preservative-impregnated or degraded timber state inhibiting value recovery (TNS Challenge 5/ VinylPlus Challenge 5).

Cast (ductile) iron recommended as an alternative to PVC water pipes[12] was formerly commonly used for this purpose. Iron is, of course, mined from the Earth's crust and its manufacture is energy-intensive. Ductile iron tends to corrode – I know this well as a ductile pipe corroded and burst under the living room of my house after around 40 years of life! – and consequently, although theoretically recoverable for recycling with its own energy intensity issues, the degraded state of end-of-life pipes limit their suitability for recycling of their remaining iron content. Although the service life of iron pipes can be extended by polymer coating, this composite form is not a fair comparison with PVC pipes that are light, flexible and so resilient to a high degree of stress and distortion and are also then recoverable and recyclable at end-of-life.

- Considerable recovery and re-melting of ductile iron take place, albeit with some loss due to degradation in service life (TNS Challenge 2/VinylPlus Challenge 1). Ductile iron has little or no additive use other than in production (TNS Challenge 4/VinylPlus Challenge 3), noting that many iron water pipe applications today use a polymer coating to prolong life but this is excluded from assessment as it is not directly relevant to comparison of pure ductile iron with PVC pipes. Few if any persistent substances, other than iron itself that is common in the Earth's crust and can be recycled for beneficial reuse, are entailed in ductile iron pipes beyond manufacturing phases (TNS Challenge 3/VinylPlus Challenge 2). Although ductile iron manufacturing is energy- and carbon-intensive, recycling captures embedded energy albeit that the weight of iron pipes also adds to carbon emissions in transport (TNS Challenge 1/VinylPlus Challenge 4). Consortia such as the International Council on Mining & Metals promote sustainable development through ten principles,[13] and ductile iron manufacturing industries are subject to continuous improvement and more stringent regulation, though there is no evidence of full value chain stewardship of ductile iron beyond product manufacture (TNS Challenge 5/VinylPlus Challenge 5).

Polyolefin plastics commonly used in cable insulation, as well as some pipe applications. Polyolefins such as polyethylene and polypropylene are highly flammable, lacking the inherent fire-quenching properties of PVC related to its chlorine content. Consequently, fire retardant additives to polyolefins have to be considered in a life cycle comparison, as does the recyclability of both types of finished polymer compound. By contrast, chlorinated polyvinyl (CPVC: PVC polymer reacted with additional chloride boosting total chlorine by mass from 57% to approximately 67%), a standard material used for sprinkler systems, is almost entirely resistant to fire complying with fire and smoke regulations such as the CEN EN 13501-1EN standard.[14] Unmodified PVC also has strong fire-quenching characteristics and is widely used in wire and cables, also matching safety standards and maintaining its beneficial properties when mechanically recycled. For polyolefins and other types of plastic, additional smoke suppressants are required in these applications. These non-PVC plastic types can potentially be recycled chemically with substantial energy inputs, though brominated flame retardants can be an obstacle to recycling. However, polyolefins also lack the chemical properties that enable PVC to be mechanically recycled with far higher energy efficiency.

- Polyolefins are carbon-intensive in terms of manufacturing energy as well as embodied carbon, which is far higher than PVC polymer which comprises 57% by weight chlorine atoms. Some recycling of polyolefins occurs, recovering carbon content though at a low rate compared to production (TNS Challenge 1/VinylPlus Challenge 4). There is substantial use of polyolefins in short-life applications which lack effective recovery at end-of-life, contributing significantly to contemporary visible marine litter accumulation, and recycling of polyolefins in other applications is at a low rate compared to production volumes (TNS Challenge 2/VinylPlus Challenge 1). The inherent flammability of polyolefin molecules means that compounds used in cable insulation as well as pipes and other applications require the incorporation of substantial concentrations of flame-suppressant additives (TNS Challenge 4/VinylPlus Challenge 3). The low level of recovery of polyolefins from end-of-life products means that the risk of persistent organic compound generation throughout societal life cycle, particularly on disposal, is far from minimal (TNS Challenge 3/VinylPlus Challenge 2). In January 2018, the Polyolefin Circular Economy Platform[15] of PlasticsEurope published a Voluntary Commitment to 2030 as part of a wider PlasticsEurope target to increase circularity and resource efficiency, comprising targets and initiatives focused on increasing reuse and recycling, preventing plastics leakage into the environment and accelerating resource efficiency, an initiative that is very welcome albeit starting from a later baseline than PVC voluntary commitments (TNS Challenge 5/VinylPlus Challenge 5).

The summary comparative assessments of the use of these four differing materials, one from PVC and the other three selected as proposed alternatives to PVC in different applications, highlight that all materials have sustainability concerns associated with their unique properties, but that they also share common challenges relating to lack of join-up along societal life cycles. Some challenges are being embraced proactively, some partially so, and others are not yet addressed. This is not a full 'level playing field' assessment, but it serves to highlight that all materials have issues requiring attention from the perspective of a range of sustainability principles and when the full societal life cycle of the products into which they are incorporated (not least at and beyond end-of-life) are taken into account. For copper in pipes, as one example, companies in the copper industry champion the material but also tend to overlook the inherent toxicity and cost of the metal whilst also misrepresenting counter-arguments "…commonly seen in the plastics green-wash",[16] and the 2023 report *The Perils of PVC Plastic Pipes*[17] by the clearly far from neutral organisation 'Beyond Plastics' misrepresents PVC whilst substantially overlooking environmental and economic issues raised by its proposed wholesale replacement with copper in the US. As another example, life cycle assessments of linoleum flooring suggest that the biological basis of much, but not all, of the inputs of raw materials means that they are automatically benign,[18,19] notwithstanding that farming is the greatest negative pressure of biodiversity, habitat loss and water use globally.[20] Additional comparative analysis on a genuinely like-for-like basis of materials suggested as alternatives to PVC or in their own right would be informative as each has its own distinctive supply chain, toxicity, product durability, maintenance requirements, non-recyclability and other challenges across life cycles influencing their relative suitability for meeting different human needs safely and efficiently.

The purpose of the analysis in the 2020 paper and its brief summary here was not to suggest that PVC is more sustainable than other materials. The rubric applied in the comparative analysis was simply based on identifying issues of concern on a challenge-by-challenge basis, and the extent to which they were being addressed on a proactive basis. Nor was the intent to dismiss or overlook very welcome and necessary progress with sustainable development associated with all of these materials. However, the comparative analysis highlights that issues of concern as well as of benefit are often highly context-specific for different materials and also in differing applications. The comparison also emphasises the need for systems thinking as all challenges are intrinsically linked, and all are also relevant across material types.

The brief comparative analysis also demonstrated that the five TNS Sustainability Challenges for PVC remain relevant 20 and more years after their initial publication, having stimulated tangible progress with sustainable development in the PVC sector. Furthermore, they also remain evidently relevant for assessment of, and application to, the sustainable development of the use of all material types. The analysis also highlights dangers inherent in automatic selection or deselection of materials in the absence of assessment of options on a 'level playing field' of sustainability principles.

6.2 A LEVEL PLAYING FIELD: ASF

As outlined earlier in this book, the Additive Sustainability Footprint (ASF) tool was developed as a collaboration between VinylPlus and The Natural Step as an adaptation of the TNS SLCA approach specifically for addressing the challenge of the sustainable use of additives (TNS Challenge 4/VinylPlus Challenge 3).

The structure of the ASF tool is outlined previously – at its core an orientation of TNS System Conditions against life cycle assessment stages supported by three 'impact questions' and four 'progress questions' – within a ten-step process largely consistent with those of the ISO 14040 standard on life cycle assessment. Much like the five TNS Sustainability Challenges for PVC, I was very mindful throughout the process of co-developing the ASF tool that it should not ultimately apply only to PVC, inadvertently reinforcing unfounded assumptions that PVC is a material apart from the many others in societal use. It was important that ASF should have generic relevance to the wide range of materials used by society. For this reason, whilst the ASF language is relevant for sustainability assessment of PVC additives across whole product value chains, in essence the tool is an adaptation of SLCA for the generic assessment of chemicals across whole societal product life cycles.

A peer-reviewed scientific paper that I published in 2022 titled *Assessment of the sustainable use of chemicals on a level playing field*[21] considered ASF alongside other commonly used chemical assessment tools, taking account of the degree to which they were based on a range of principles relevant to the sustainable use of chemicals over whole product life cycles. These broader principles of sustainable use included addressing multiple dimensions of sustainability beyond simple chemical characteristics; a foundation in science; consideration of life cycle risk rather than simply intrinsic chemical properties; positive contributions to meeting human needs; open access to tools; whether they are statutory; and that they have been subjected to peer review.

The 2022 paper observed that misleading conclusions can readily be drawn when findings from the application of tools developed for specific purposes are extrapolated to wider conclusions about overall sustainable use. This is particularly pertinent for tools with a narrow focus on intrinsic chemistry rather than evaluation of their use within the context of whole product life cycles, and also in terms of whether they take account of benefits to society amongst wider socio-economic factors. At present, only the ASF approach accounts for how the use of substances contributes beneficially to the meeting human needs: the 'lost half' of sustainable development as discussed previously. Most other chemical assessment approaches are substantially based on 'intrinsic properties', and hence potential for hazard in isolation from wider life cycle exposure and risk assessment. Orientation of chemical assessment approaches against principles relevant to sustainable use in the 2022 paper is summarised in Table 6.1, using a 'traffic lights' colour coding signifying whether the sustainability-relevant principle is fully met (green), partially met (amber) or not met (red).

At the risk of sounding like a stuck record (albeit that vinyl records are a popular use of PVC that is seeing something of a renaissance!) the sustainability contribution of a chemical cannot be determined by narrow assessments based on intrinsic properties alone, nor simplistic judgements about what is 'natural', 'good' or other such lazy and generally misleading tags. Another scientific paper published by other authors in 2020 concurred that, whilst chemicals policies have generated a wide range of regulations aimed at limiting damage to the environment and human health, "Most instruments are reactive and fragmented".[22]

Some 'natural' or at least nature-based substances can have adverse health implications, which is unsurprisingly when we think back to the description early in this book about the many toxins, venoms, deterrents and other biologically active substances evolved by living organisms for defensive and other purposes. Furthermore, inherently biodegradable natural substances,

TABLE 6.1   Summary of chemical assessment systems/approaches in terms of their coverage of criteria relevant to sustainable use, emphasised by colour coding

	FULL DIMENSIONS OF SUSTAINABLE DEVELOPMENT	TRANSPARENTLY SCIENCE-BASED	BASED ON FULL ARTICLE LIFE CYCLE RISK (RATHER THAN POTENTIAL HAZARD ALONE)	RECOGNISES POSITIVE CONTRIBUTIONS TO MEETING HUMAN NEEDS	OPEN ACCESS	FREE TO USE (ALBEIT WITH GUIDANCE AND EXTERNAL AUDITING)	APPLICABLE ACROSS PRODUCTS/MATERIALS	STATUTORY	PEER REVIEWED IN SCIENCE LITERATURE
Life Cycle Assessment (LCA)	NO	YES	Partially	NO	Partially	Partially	YES	NO	YES
Environmental Product Declaration (EPD)	NO	YES	Partially	NO	Partially	YES	YES	NO	YES
Product Environmental Footprint (PEF)	NO	YES	Partially	NO	Partially	YES	YES	NO	YES
EU REACH	NO	YES	NO	NO	YES	YES	YES	YES	YES
SciveraLENS®	NO	YES	Partially	NO	NO	NO	YES	NO	NO
Greensuite®	NO	YES	Partially	NO	NO	NO	YES	NO	YES
GreenScreen List Translator™	NO	YES	NO	NO	NO	NO	YES	NO	Partially
GreenWERKS	NO	YES	NO	NO	NO	NO	YES	NO	NO
Green Chemistry and Commerce Council (GC3) Retailer Database	NO	YES	NO	NO	Partially	Partially	YES	NO	NO
OECD Substitution and Alternatives Assessment Toolbox (SAAT)	NO	YES	NO	NO	Partially	Partially	YES	NO	NO
ECHA Plastic Additives Initiative	NO	YES	Partially	NO	YES	YES	Partially	NO	NO
Cradle to Cradle	YES	YES	YES	NO	NO	NO	YES	NO	YES

(Continued)

TABLE 6.1 (*Continued*) Summary of chemical assessment systems/approaches in terms of their coverage of criteria relevant to sustainable use, emphasised by colour coding

	FULL DIMENSIONS OF SUSTAINABLE DEVELOPMENT	TRANSPARENTLY SCIENCE-BASED	BASED ON FULL ARTICLE LIFE CYCLE RISK (RATHER THAN POTENTIAL HAZARD ALONE)	RECOGNISES POSITIVE CONTRIBUTIONS TO MEETING HUMAN NEEDS	OPEN ACCESS	FREE TO USE (ALBEIT WITH GUIDANCE AND EXTERNAL AUDITING)	APPLICABLE ACROSS PRODUCTS/MATERIALS	STATUTORY	PEER REVIEWED IN SCIENCE LITERATURE
Additive Sustainability Footprint (ASF)	YES	YES	YES	YES	YES	YES	YES	NO	YES
Ecovadis	YES	YES	NO	NO	NO	NO	YES	NO	NO
Carbon Handprint	NO	YES	YES	NO	YES	YES	YES	NO	YES
Material flow cost accounting (MFCA)	NO	YES	Partially	NO	NO	NO	Partially	NO	YES
GRI 301: Materials	NO	YES	NO	NO	YES	YES	YES	NO	NO

GREEN: Yes, fully meets criterion; AMBER: Partially meets criterion; RED: No, does not meet criterion.

ostensibly 'greener' without critical appraisal, may only offer short service lives or else require inputs of biocides or other preservatives during manufacture or use to maintain or prolong service life with associated sustainability implications, beyond which they may not be recoverable.

By contrast, synthetic substances such as plastics – often demonised and sometimes for good reasons where responsible manufacture and sound life cycle stewardship are lacking – are widely used for medical benefits due to their inert and adaptable properties, and they also offer very long service life with low or no inputs before being recycled in durable applications such as the 69% of PVC that is used in construction and building products.

As previously described, intrinsically 'safe' chemicals can be unsafe in use. By contrast, intrinsically hazardous substances may be entirely consumed during contained material production, or immobilised in products and then recovered safely and efficiently through circular use with no risk to the environment or human health.

These factors explain why it is so important to consider the use of materials across the whole societal life cycles of the products into which they are incorporated. It is important too to address dimensions of sustainability beyond negative potential toxic effects, and to include dimensions such as positive contributions to meeting human needs. Social and health implications from supply chains through manufacture and use and to end-of-life and beyond are also relevant to sustainability performance.

Of the chemical assessment systems/approaches addressed in terms of their coverage of criteria relevant to sustainable use in the 2022 'level playing field' paper, only ASF – an application of the SLCA approach to address the societal use of chemicals – meets all criteria, with the singular exception of inclusion as a statutory approach. It is one of only a few assessment approaches addressing wider dimensions of sustainable development and also addressing societal life cycles. Significantly, ASF is the only approach found to recognise positive contributions to meeting human needs, which are overlooked in other approaches that focus narrowly on potential negative outcomes. ASF is therefore the only assessment approach addressing the important 'lost half' of sustainable development of how the use of materials can contribute to meeting diverse societal needs in the face of resource depletion and other mounting sustainability pressures. ASF is also open and free to use, albeit that its implementation benefits from bespoke training, and is also subject to peer review. It is for these reasons that ASF provides a comprehensive basis for assessment of the sustainable use of substances. It is, in other words, explicitly designed for assessment of the sustainable use of chemicals across whole product life cycles on a level playing field.

The 2022 paper *Assessment of the sustainable use of chemicals on a level playing field*[23] consequently proceeded to ground and demonstrate the wider utility of ASF by applying it to the use of a range of qualitatively differing material types across their whole final product life cycles. The ASF-assessed comparative substances in this 2022 study were as follows:

- **Metal-based stabiliser additives in PVC compounds** used in a generic EU-wide window profile formulation, broader findings of which have already been noted in Table 5.3 but that are summarised in Table 6.2;
- Use of **brominated flame retardants in polyolefins** used in cables and other applications;
- **Timber**, commonly regarded as an eco-friendly material due to its biological origins, ASF-assessed here in its use in window profiles; and
- Use of **mined cobalt**, for example in production of some solar panels.

These brief comparative ASF analyses highlight illustrative examples of pollution-related, resource depletion and ethical issues, addressing both challenges and benefits by TNS System Condition at different stages in product life cycles. Findings are summarised in Table 6.2, which also shows potential positive (☺) and negative (☹) implications and potential mitigation measures (↑) throughout final product life cycles, with some further summary points added in addition to those in the peer-reviewed paper.

Highlights of some of these analyses include:

- Polyolefin-based plastics are often presented as a more benign alternative to PVC as they do not contain chlorine. However, the chlorine atoms lend PVC self-quenching properties, whereas polyolefins are highly flammable requiring additions of flame retardants, including hazardous brominated flame retardants. These flame retardants can substantially influence the sustainability footprint of polyolefin compounds, particularly brominated flame retardants that are hard to recycle.
- Timber is commonly regarded as an eco-friendly material in window profiles and other uses due to its biological origins. However, ASF analysis taking account of the total product life cycle highlights issues and potential mitigation measures associated with sourcing, limited service life, inputs of biocidal/preservative chemicals in product manufacture and/or use, and their potential to inhibit benign breakdown, recycling or further beneficial use of the timber.
- Cobalt is an important constituent of some renewable energy applications yielding substantial benefits across TNS System Conditions during product life. However, ASF analysis highlights both ethical and pollution issues particularly at the raw material extraction phase, as well as likely pollution issues at product end-of-life, with further issues requiring management in intervening life cycle stages.
- Metal stabilisers used in PVC window profiles have some issues requiring innovation throughout the product life cycle, as noted in the previous detailed analysis, yet the resilience and long service life conferred by use of stabilisers results in low material, energy and staff time inputs during an extended service life, then with the potential for material value recovery through controlled loop recycling at end-of-life for further cycles of beneficial use in products.

These illustrative ASF assessments demonstrate that many of the challenges associated with the use of materials, when assessed on a systemic basis, are common to many or perhaps all materials resultant from unsustainable societal material use habits. This observation underlines the dangers of assumptions that some materials are inherently 'good' whilst others are 'bad' when assessments are divorced from consideration of stewardship across whole product

TABLE 6.2 Illustrative examples of potential positive (☺) and negative (☹) sustainability implications and mitigation measures (↑) at specific points in final product life cycles in the use of selected chemicals (with more complete summary of metal-based PVC additives for comparison)

	LIFE CYCLE STAGES					
	RAW MATERIALS ACQUISITION	ADDITIVE PRODUCTION	PACKAGING AND DISTRIBUTION	COMPOUNDING/ CONVERTING	PRODUCT USE/ MAINTENANCE	POST-USE MANAGEMENT
Use of metal-based stabiliser additives in PVC window profiles	☹ Risks of metal release (SC1), persistent emissions (SC2), physical damage (SC3) and associated health and ethical impacts (SC4) can be ↑ contained by using recycled metal or best practice in virgin extraction	☹ Risks of metal release (SC1), persistent emissions (SC2), physical damage (SC3) and associated health impacts (SC4) can be ↑ contained by controlled manufacturing with post-industrial waste recycling	☹ Risks of metal release (SC1), persistent emissions (SC2), physical damage (SC3) and associated health impacts (SC4) can be ↑ contained by efficient and tightly controlled transport	☹ Risks of metal release (SC1), persistent emissions (SC2), physical damage (SC3) and associated health impacts (SC4) can be ↑ contained by controlled compounding with post-industrial waste recycling or safe disposal	☺ The resilience of stabilised PVC gives long service life with low maintenance, delivering societal value with low/no inputs on emissions (SC1, SC3, SC4)	☹ Risks of metal release (SC1), persistent emissions (SC2), physical damage (SC3) and associated health impacts (SC4) and resource wastage from incautious disposal can be ↑ contained by controlled loop recycling
Use of brominated	☹ Mining of chemicals (SC1) and bromine extraction (SC2)	☹ When based on mined petrochemicals (SC1), can lead to	☹ Risks of loss of brominated flame retardants in packaging	☹ Risks of loss of brominated flame retardants in compounding	☺ Flame retardants reduce fire-related human health risks (SC4)	☹ Health risks (SC4) and ecological harm including

(Continued)

TABLE 6.2 (Continued) Illustrative examples of potential positive (☺) and negative (☹) sustainability implications and mitigation measures (↑) at specific points in final product life cycles in the use of selected chemicals (with more complete summary of metal-based PVC additives for comparison)

	LIFE CYCLE STAGES					
	RAW MATERIALS ACQUISITION	ADDITIVE PRODUCTION	PACKAGING AND DISTRIBUTION	COMPOUNDING/ CONVERTING	PRODUCT USE/ MAINTENANCE	POST-USE MANAGEMENT
flame retardants in polyolefins used in cables	used in manufacture of brominated flame retardants can disrupt landscapes, brine lagoons and aquifers (SC3) though, ↑ However, bio-based organic content potentially reduces this problem ☺ However, biomass-based production can potentially compete with food security[24] (SC4), requiring cautious auditing of supply chains	emission and accumulation of climate-active gases (SC2) though, ↑ bio-based manufacture potentially reduces this problem ☺ Potential for systematic accumulation of fugitive bromine and bromine-containing compounds in ecosystems[25] including long-range transport and bioaccumulation[26] (SC2) and in human tissues[27] posing health risks (SC4)	and transport (SC2) with possible health risks (SC4) can be ↑ contained by efficient and tightly controlled transport	or converting (SC2) with possible health risks (SC4) can be ↑ contained by efficient and tightly controlled processes	☹ Health risks (SC4) and ecological harm including bioaccumulation (SC2) can occur as brominated flame retardants are known to leach out of plastic products and may do so in simulated use conditions[28] and in landfill[29]	bioaccumulation (SC2) can occur if brominated flame retardants are released during product disposal, as they are found to leach out in water, methanol and humic substances However, ↑ excessive leaching can be averted by controlled disposal ☺ Nonetheless, technical obstacles remain for the physical recycling of plastics containing brominated flame retardants, though some less efficient chemical recycling routes are available[30]

(Continued)

TABLE 6.2 (Continued) Illustrative examples of potential positive (☺) and negative (☹) sustainability implications and mitigation measures (↑) at specific points in final product life cycles in the use of selected chemicals (with more complete summary of metal-based PVC additives for comparison)

	LIFE CYCLE STAGES					
	RAW MATERIALS ACQUISITION	*ADDITIVE PRODUCTION*	*PACKAGING AND DISTRIBUTION*	*COMPOUNDING/ CONVERTING*	*PRODUCT USE/ MAINTENANCE*	*POST-USE MANAGEMENT*
Timber used in window profiles including necessary preservative additions to maintain service life	☹ Non-sustainable timber extraction can damage ecosystems also including soil and aquifer (SC3), mobilise sequestered carbon (SC1) and impinge of forest-dweller rights (SC4) However, ↑ sustainable forestry practices can limit ecological harm (SC3) and potentially be carbon neutral (SC1) and ethical (SC4)	☹ Manufacture of finished timber entails use of energy generally from carbon-based source (SC1) and can generate emissions (SC2), though ↑ these can be mitigated by best practice ☹ Addition of preservatives (SC2) during manufacture or during window profile use can generate pollutants (SC2) that may be hazardous to health (SC4), though ↑ can be mitigated by best practice and lower-toxicity additives	☹ Fossil fuel use in transport (SC1) with risk of persistent emissions (SC2), physical damage (SC3) and associated health impacts (SC4) can be ↑ contained by efficient and tightly controlled transport	Window profile production from finished timber is generally safe and low in hazard if best practice if followed	☹ Periodic inputs of biocidal timber treatment (SC2) are required to extend inherently biodegradable product life, or wood product requires frequent replacement, both combining to produce a shorter service life and societal value (SC4) ☺ Well-manufactured wooden window profiles provide comfort, sound-proofing and thermal insulation (SC4, SC1)	☹ Biocidal timber treatment can inhibit recycling at end-of-life, leading to linear resource use (SC1, SC2, SC3) denying beneficial reuse with lost value to society (SC4) This may be ↑ mitigated to some extent by best practice incineration with heat recovery, though this is the lowest tier of circular economic use

(Continued)

TABLE 6.2 (Continued) Illustrative examples of potential positive (☺) and negative (☹) sustainability implications and mitigation measures (↑) at specific points in final product life cycles in the use of selected chemicals (with more complete summary of metal-based PVC additives for comparison)

	LIFE CYCLE STAGES					
	RAW MATERIALS ACQUISITION	ADDITIVE PRODUCTION	PACKAGING AND DISTRIBUTION	COMPOUNDING/ CONVERTING	PRODUCT USE/ MAINTENANCE	POST-USE MANAGEMENT
Use of mined cobalt, for example in production of some solar panels	☹ As cobalt is a scarce metal, it may systematically accumulate in nature during incautious extraction and purification (SC1), and open cast mining is damaging to ecosystems including groundwater (SC3)	☹ Risks release of cobalt and co-contaminants (SC1), persistent emissions (SC2), physical damage (SC3) and associated health impacts (SC4) can be ↑ reduced by controlled manufacturing	☹ Risks of cobalt release (SC1), persistent emissions (SC2), physical damage (SC3) and associated health impacts (SC4) can be ↑ contained by efficient and tightly controlled transport	☹ Risks of cobalt release (SC1), persistent emissions (SC2), physical damage (SC3) and associated health impacts (SC4) can be ↑ contained by controlled compounding with post-industrial waste recycling or safe disposal	☺ Solar panel applications including mined cobalt can be long-lived yield high societal value (SC4) with significant savings in carbon emissions relative to fossil fuel-based generation (SC2)	☹ Cobalt is a scarce metal in nature that may systematically accumulate in nature with incautious disposal (SC1) However, ↑ controlled loop recovery can retain material value and avert health risks (SC4) as well as pollution (SC1)

(Continued)

TABLE 6.2 (Continued)   Illustrative examples of potential positive (☺) and negative (☹) sustainability implications and mitigation measures (↑) at specific points in final product life cycles in the use of selected chemicals (with more complete summary of metal-based PVC additives for comparison)

	LIFE CYCLE STAGES				
RAW MATERIALS ACQUISITION	ADDITIVE PRODUCTION	PACKAGING AND DISTRIBUTION	COMPOUNDING/ CONVERTING	PRODUCT USE/ MAINTENANCE	POST-USE MANAGEMENT
☹ Cobalt mining in some regions is associated with human rights abuses, funding repressive regimes, and creating geopolitical inequities[31,32] (SC4). Congo accounts for about 70% of world cobalt mine as a by-product of copper or nickel[33] ↑ Responsible sourcing can minimise risks					

life cycles, from raw material sourcing through manufacturing, product life, and at and beyond product end-of-life. Issues associated with substance use at all life cycle stages can exert substantial influence on overall sustainability characteristics.

Substances commonly portrayed as problematic can deliver long service lives optimising delivery of needs per unit of resource due to their durability and resistance, as well as being amenable to value recovery (both material and embodied energy) through recycling at the end of the useful life of products. These comparative ASF-based analyses highlight that, although important considerations apply to some life cycle and TNS System Condition stages, the intrinsic chemical properties of substances are only part of the wider breadth of issues germane to wider-ranging sustainability assessment. Significantly, these wider considerations include the delivery of optimal societal value through addressing needs in the safest and most eco-efficient manner when considered in the context of product life cycles.

It is therefore essential to make wise judgements about material sustainability performance on a science-based and systemically framed 'level playing field', which also serves to highlight 'hotspots' requiring sustainable innovation throughout the entire product life cycle.

6.3 NECESSARY EVOLUTIONS IN REGULATION, ASSUMPTIONS AND COMMON UNDERSTANDING

These observations have major implications for the regulation of chemicals. As we have seen, the vast bulk of established tools supporting regulations, including such continental-scale regulatory processes as REACH in the European Union, focus narrowly on intrinsic chemical properties (potential hazard without full account of actual risk). They do so with a limited disciplinary focus largely based on potential for harm, also failing to account of the beneficial contributions of the use of substances to help meet human needs.

There are also deeper questions in terms of even-handed and consistent regulation of chemicals used for different purposes. In Europe, for example, the EU REACH regulations pertain to bulk chemical use, but exclude pharmaceuticals and pesticides each of which is covered by discrete regulations that differ in assessment protocols. In the US, many substances in widespread use before the Environmental Protection Agency (EPA) was established in

1970, and particularly when its remit was extended to controls on substances in use in 1974, were tacitly authorised under a 'grandparenting' approach lacking systemic scrutiny. The 'grandparenting' of many substances, including harmful and persistent substances often referred to as 'forever chemicals', has raised public concern including for example Teflon (polytetrafluoroethylene, or PTFE, patented by the DuPont company in 1945) and perfluorooctanoic acid (PFOA, or C8) that achieved notoriety through legal cases in the US.[34]

The basis of chemical regulation therefore needs wholesale evolution and consistency – between substance applications, between nations and taking a life cycle risk approach – if it is to become more relevant to driving progress and innovation relevant to the wider scope of sustainable development. There is an urgent need for regulations to embrace wider sustainability principles, and to evolve from over-simplistic – albeit more simply implemented – potential hazard-based assessment into a risk-based approach taking account of whole product life cycles. The ASF approach appears to offer a science-based and transparent framework for accomplishing this goal, and has already been peer-reviewed and applied in practice. Whilst the full ASF process is onerous, it can also be applied as a 'light touch' for systemic screening of issues of potential concern (as for example in comparative examples in this chapter). We will return to regulatory evolution and practical implementation in the following chapter.

These wider perspectives also need to influence established assumptions in business and society, better to refocus on the pressing need to rethink material choice and innovation around what best helps people meet various of their needs in the safest, most ethical and efficient manner taking account of whole societal life cycles. This perspective is far from academic or altruistic; a wise business will be vigilant about pre-empting future risks, liabilities, restrictions, negative publicity and customer demands, as well as seeking wise and profitable investments in novel opportunities in a world increasingly shaped by 'The Funnel' effect of sustainability pressures. A wise business needs to make choices and investments informed by a systemic level playing field.

NOTES

1   Greenpeace. (1998). *Alternatives to PVC*. Greenpeace.
2   CEC. (2000). *Green Paper: Environmental Issues of PVC*. Commission of the European Communities (CEC), Brussels. [Online.] https://ec.europa.eu/environment/pdf/waste/pvc/en.pdf, accessed 13 April 2023.

3 NCBE. (1997). *Summary Report for PVC Retail Working Group.* National Centre for Business and Ecology, Manchester.

4 European Commission. (2004). *Life Cycle Assessment of PVC and of principal competing materials.* European Commission. [Online.] https://vdocuments.mx/pvc-final-report-lca-en.html?page=1, accessed 29 April 2023.

5 BPF. (2019). *Designing in a Circular Economy with PVC.* British Plastics Federation. (BPF), 31 May 2019. [Online.] https://www.bpf.co.uk/Sustainability/designing-in-a-circular-economy-with-pvc.aspx, accessed 14 April 2023.

6 Johnsen, T. (2023). Is PVC the answer to a circular economy in healthcare? *Medical Plastics News,* 10 March 2023. [Online.] https://www.medicalplastic-snews.com/medical-plastics-industry-insights/medical-plastics-sustainability-insights/is-pvc-the-answer-to-a-circular-economy-in-healthcare/, accessed 14 April 2023.

7 Everard, M. (2020). Twenty years of the Polyvinyl Chloride (PVC) sustainability challenges. *Journal of Vinyl and Additive Technology.* 26(3), pp. 390–402. DOI: https://doi.org/10.1002/vnl.21754.

8 Greenpeace. (n.d.). *Your Choice of Window Could Seriously Harm the Planet.* Cited in Green Building Store. (2019). *Alternatives to PVC.* Green Building Store. [Online.] https://www.greenbuildingstore.co.uk/information-hub/alternatives-to-pvc/, accessed 14 March 2023.

9 Thompson C. (2005). *Window of Opportunity: The Environmental and Economic Benefits of Specifying Timber Window Frames.* WWF-UK, Godalming. [Online]. https://www.wwf.org.uk/sites/default/files/2017-06/windows_0305.pdf, accessed 14 March 2023.

10 UNEP. (n.d.). *Facts and Figures about the Minamata Convention.* UN Environment Programme (UNEP). [Online.] https://mercuryconvention.org/en, accessed 17 March 2023.

11 UNECE. (n.d.). *Gothenburg Protocol.* UNECE. [Online.] https://unece.org/gothenburg-protocol, accessed 17 March 2023.

12 Mother Earth Living. (2004). *Pipe Dreams: Alternatives to PVC Plumbing Pipes.* Mother Earth Living. [Online]. https://www.motherearthliving.com/home-products/nh-builders-corner, accessed 14 March 2023.

13 ICMM. (n.d.). *Our principles.* International Council on Mining & Metals (ICMM). [Online.] https://www.icmm.com/en-gb/our-principles, accessed 17 March 2023.

14 CEN. (2018). *CEN - EN 13501-1: Fire Classification of Construction Products and Building Elements - Part 1: Classification Using Data from Reaction to Fire Tests.* European Committee for Standardization (CEN). [Online.] https://standards.globalspec.com/std/13162977/EN%2013501-1, accessed 11 April 2023.

15 PCEP. (2018). *PCEP Polyolefin Circular Economy Platform.* PlasticsEurope. [Online.] https://pcep.eu/, accessed 17 March 2023.

16 CuSPUK. (2023). *On the CuSP of a More Sustainable Future.* Copper Sustainability Partnership (CuSP). [Online.] https://www.cuspuk.com/, accessed 13 May 2023.

17 Beyond Plastics. (2023). *The Perils of PVC Plastic Pipes*. Beyond Plastics, April 2023. [Online.] https://www.beyondplastics.org/publications/perils-of-pvc-pipes, accessed 14 May 2023.

18 Jönsson, Å., Tillman, A.-M. and Svensson, T. (1997). Life cycle assessment of flooring materials: case study. *Building and Environment*, 32(3), pp. 245–255. DOI: https://doi.org/10.1016/S0360-1323(96)00052-2.

19 Wilson, A. (n.d.). Linoleum: the all-natural flooring alternative. *Environmental Building News*, 7(9). [Online.] https://www.buildinggreen.com/feature/linoleum-all-natural-flooring-alternative, accessed 13 May 2023.

20 IPBES. (2019). *Global Assessment Report on Biodiversity and Ecosystem Services*. Intergovernmental Panel on Biodiversity and Ecosystem Services (IPBES). [Online.] https://ipbes.net/global-assessment, accessed 13 May 2023.

21 Everard, M. (2022). Assessment of the sustainable use of chemicals on a level playing field. *Integrated Environmental Assessment and Management*, 19(4), pp. 1131–1146. DOI: https://doi.org/10.1002/ieam.4723.

22 Collins, C., Depledge, M., Fraser, R., Johnson, A., Hutchison, G., Matthiessen, P., Murphy, R., Owens, S. and Sumpter, J. (2020). Key actions for a sustainable chemicals policy. *Environment International*, 137, p. 105463. DOI: https://doi.org/10.1016/j.envint.2020.105463.

23 Everard, M. (2022). Assessment of the sustainable use of chemicals on a level playing field. *Integrated Environmental Assessment and Management*. DOI: https://doi.org/10.1002/ieam.4723.

24 Chen, X. and Önal, H. (2016). Renewable Chen energy policies and competition for biomass: implications for land use, food prices, and processing industry. *Energy Policy*, 92, pp. 270–278. DOI: https://doi.org/10.1016/j.enpol.2016.02.022.

25 Domínguez, A.A., Law, R.J., Herzke, D. and de Boer, J. (2010). Bioaccumulation of Brominated Flame Retardants. In: Eljarrat, E., Barceló, D. (eds) *Brominated Flame Retardants. The Handbook of Environmental Chemistry*, 16. Springer, Berlin, Heidelberg. DOI: https://doi.org/10.1007/698_2010_95.

26 Lee, H-J. and Kwon, J-H. (2020). Chapter Six - Persistence and Bioaccumulation Potential of Alternative Brominated Flame Retardants. In: Oh, J-E. (ed.) *Comprehensive Analytical Chemistry*. Elsevier, 88, pp. 191–214. DOI: https://doi.org/10.1016/bs.coac.2019.10.005.

27 Weijs, L., Dirtu, A.C., Malarvannan, G. and Covaci, A. (2015). Chapter 14 - Bioaccumulation and Biotransformation of Brominated Flame Retardants. In: Zeng, E.Y. (ed) *Comprehensive Analytical Chemistry*, 67, pp. 433–491. DOI: https://doi.org/10.1016/B978-0-444-63299-9.00014-4.

28 Bohlin-Nizzetto, P. (2023). *Content and Migration of Chemical Additives from Indoor Consumer Plastic Products*. Norwegian Institute for Air Research (NILU), NILU report 6/2023.

29 Kim, Y-J., Osako, M. and Sakai, S-i. (2006). Leaching characteristics of polybrominated diphenyl ethers (PBDEs) from flame-retardant plastics. *Chemosphere*, 65(3), pp. 506–513. DOI: https://doi.org/10.1016/j.chemosphere.2006.01.019.

30 Ma, C., Yu, J., Wang, B., Song, Z., Xiang, J., Hu, S., Su, S. and Sun, L. (2016). Chemical recycling of brominated flame retarded plastics from e-waste for clean fuels production: a review. *Renewable and Sustainable Energy Reviews*, 61, pp. 433–450. DOI: https://doi.org/10.1016/j.rser.2016.04.020.

31 Sovacool, B.K. (2019). The precarious political economy of cobalt: Balancing prosperity, poverty, and brutality in artisanal and industrial mining in the Democratic Republic of the Congo. *The Extractive Industries and Society*, 6(3), pp. 915–939. DOI: https://doi.org/10.1016/j.exis.2019.05.018.
32 Karlsson, C-J. and Zimmer, K. (2020). Green energy's dirty side effects: The global transition to renewables could lead to human rights abuses and risks exacerbating inequalities between the West and the developing world. *Foreign Policy*, 18 June 2020. [Online.] https://foreignpolicy.com/2020/06/18/green-energy-dirty-side-effects-renewable-transition-climate-change-cobalt-mining-human-rights-inequality/, accessed 31March 2023.
33 USGS. (2023). *Cobalt*. US Geological Survey (USGS). [Online.] https://pubs.usgs.gov/periodicals/mcs2023/mcs2023-cobalt.pdf, accessed 15 May 2023.
34 EcoWatch. (2016). *Teflon's Toxic Legacy: DuPont Knew for Decades It Was Contaminating Water Supplies*. EcoWatch, 4 January 2026. [Online.] https://www.ecowatch.com/teflons-toxic-legacy-dupont-knew-for-decades-it-was-contaminating-wate-1882142514.html, accessed 29 April 2023.

Sustainability and the Purpose of Business and Regulation

7

Business is about making money. Without turning a profit, the business will fold. That is how the market works. We take that as a given, but there is far more to business and the workings of the globally dominant capitalist model than that: historically, in the present and into the future.

7.1 THE BUSINESS OF BUSINESS

Businesses exist because we in the developed world, and increasingly where contemporary market logic influences practices beyond that, have chosen a capitalist model as the primary means for converting natural and other resources into useful products.[1] The outset of the European Industrial Revolution saw establishment of the fundamentals of modern market economics and business assumptions about resource flows. This revolution also spawned an era of geographical prospecting and empire-building, the world seemingly boundless in terms of fresh resources as forests, mineral deposits and sources of energy-dense fuels such as coal became progressively depleted at home. It is then unsurprising that ideas such as there being environmental limits to natural resources, or of nature's capacities for waste assimilation being finite, were not embedded in the foundations of assumptions and habits developed when the global and largely undeveloped human population stood at around 77 million

DOI: 10.1201/9781003453949-7

(less than one-hundredth of the current human population of 8 billion). In fact, quite the reverse was true: nature seemed boundless but human numbers were limiting for industrialised progress, both in terms of labour serving the new factories as well as consumers of the products that they generated. Mechanisation of agriculture freed up labour to migrate to the burgeoning cities that offered employment and also unrivalled material quality of life and prospects to many in society. Meanwhile, the captains of new industry enjoyed unprecedented wealth, some of which was invested by philanthropy into social enterprises such as planned urban development, schools and public libraries from which we still benefit today. Business, in essence, was about profitably serving human needs through the conversion of natural and human resources into products under a model framed by the prevalent world view. The capitalist model has since become perhaps the most pervasive ideology that the world has ever seen.[2]

As industry expanded in range and material throughput, pollution issues were to come to light as the 'footprint' of activities began to exceed formerly overlooked limits to environmental capacities. Retrospective regulation began to be implemented. The world's first industrially focused legislation, The Alkali Act 1863 enacted by the Parliament of the United Kingdom,[3] appointed inspectors to curb aerial discharges of muriatic acid (hydrochloric acid) gas from alkali works. The Alkali Act was later amended and extended to cover other industrial pollutants. Readers will be aware of much of the vast raft of evolving legislation that has since been implemented, nationally and internationally, to control impacts of industrial processes. This has largely occurred on a reactive basis as the need to protect human health and environmental damage became evident. We can add to this a canon of health and safety legislation, and also protections on employee and wider human rights that were also far from prevalent in a formerly less ethically aware age.

Beyond these restrictive forces, the freedoms of business also experienced progressive commercial liberalisation. The neoliberalisation of business was substantially accelerated by the work of US economist Milton Friedman. Friedman was a politically influential advocate of free-market capitalism, liberated from government intervention and deregulation amongst a suite of other policies rejecting state control via fiscal policy, collectively termed as 'monetarism'. Friedman won a Nobel Prize in 1976 for his work, which had significant influence on US policy under the presidency of Ronald Reagan and in the UK during Margaret Thatcher's tenure as Prime Minister. The net result was to boost the freedoms, profits and political influence of multinational corporations and, in so doing, to reshape modern capitalism at a pan-global scale.

The term 'Greed is good' gained media profile throughout the 1980s as wealthy investors became wealthier through investments in markets yielding short-term returns. A consequence of neoliberalisation is that the role of business was seen as investment to make more money that, in a deregulated state, meant that environmental and social concerns became perceived as costs and constraints, and were serially marginalised as the rich grew richer.

This monetarist take-over was possible due to the combination of capitalism and liberal democracy in the post-Second World War era when growth to address global recession was seen as vital to drag people out of poverty and to reconstruct national economies.[4] In this regard, the policy worked by allowing the world overall to get richer. However, beyond a point of recovery, it was then to foster a model of globalisation disproportionately enabling the richer world to extend its advantages at cost to the less privileged world, also widening divisions within nations between asset-owning classes expecting returns on investment and the poorer sectors upon whom fell the burden of interest payments to service loans from the already wealthy. It is this achievement of the apotheosis of the monetarist model in a world where the power of the former Soviet Union was also crumbling along with its rather different world view that led the American political scientist Francis Fukuyama to declare that we had reached the 'end of history'[5].

This monetarist triumph was a nadir for concern for environmental and social consequences, with market regulations effectively insulating profit-taking from the costs and consequences of the means by which profit is generated. This situation perpetuates today, for example with most of us who own bank accounts having no idea nor legal culpability for activities generating profits on our investments. Ironically, these monetarist shifts occurred in an era of heightening rhetoric and acceptance of the concept of sustainable development.

Were multinational agreements, such as those under the 1992 UN 'Rio Conference' (the United Nations Conference on Environment and Development, Rio de Janeiro, Brazil, 3–14 June 1992) and ensuing national declarations of commitment to sustainable development, to be fully be believed and enacted, we would already be living in a world where economic strategies were balanced with environmental and social progress. The reality lags far behind the rhetoric, posing real and possibly existential, as well as commercial, risks for our collective future. Even as late as 2013, UK Prime Minister David Cameron was reported via sources in his own Conservative political party to have ordered aides to "…get rid of all the green crap" from energy bills in a drive to bring down costs, abandoning a hollow promise to run the greenest government ever.[6] A decade on, the same political regime was justifiably accused of "Giving up on waterways" as Ministers scrapped anti-pollution housebuilding rules imposed under European rulings to protect vulnerable nature conservation sites as 'unnecessary' obstacles to profitable housebuilders.[7] The influence of narrow monetarist thinking still runs deep in the minds of many political elites, and is embedded deeply in many business assumptions and norms as well as shareholder expectations.

Nonetheless, various initiatives, technological advances, commercial realities, legislative changes and, in some cases, voluntary commitments have regained some of losses incurred under the monetarist-dominated era, at least in some global regions. These progressive steps include the rise of Corporate Societal Responsibility (CSR), whether legally mandated or elective, under which corporate profit is reported alongside environmental and social impacts and mitigations. And then there is the dawning awareness that supply chain resilience is of growing importance in an increasingly crowded and contested world where novel resources cannot be assumed once a principal source is depleted, and with businesses also vulnerable to factors such as droughts, crop failures, flooding, insurance costs, damage from extreme weather events and civil unrest. This awareness has spawned novel independently audited market-based value chain stewardship models involving multiple stakeholders, such as the *Forestry Stewardship Council* (FSC) scheme effective since 1994 for assuring forest products and the *Marine Stewardship Council* (MSC) instituted in 1999 to address the sustainability of fisheries, both discussed previously in this book. Add to this the pervasion of media and near-universal access to the internet, and corporations also realise that they are under constant digital scrutiny by an increasingly environmentally literature and tech-savvy public, with disclosure of 'dirty secrets' having potentially serious impacts on customer and investor confidence and possibly regulatory attention. There is, in short, growing recognition that a proactive stance towards meeting human needs ethically, responsibly, safely and efficiently can be the basis of good business as sustainability pressures increasingly impinge on former freedoms but also generate new opportunities.

And so, the wheel turns from the early Industrial Revolution focus of business serving society's needs by converting resources into useful products, towards a more widely informed ethos amongst enlightened businesses of serving society's needs in the most sustainable manner possible. More has to be done to embed the values of natural, human and social capitals into the generation of financial capital, and indeed to restore socio-ecological wealth that has become hugely degraded over recent centuries. Despite its many imperfections, the reality is that the capitalist model has permeated not only more of the world, but has deepened its political roots. So, this is the model that we have to work with for business to best serve its vital societal roles.

It is clear from the preceding pages that sectors of the PVC industry have taken to heart the reality that commitment to progress with sustainable development is not only good business but is essential to position it for the future. Businesses need to sell products and generate profits to remain viable, but reframing corporate purpose as meeting societal needs in socially and environmentally optimal ways as a contribution to a better world is far more motivating that a narrow and arguably selfish profit-centred world view. And, in a world where social and environmental factors mount in importance as sustainable development pressures amplify, future good, de-risked and profitable business can be conflated with improving sustainable practice in the meeting of human needs.

7.2 THE BUSINESS OF REGULATION

As we have seen in previous chapters, the focus of a great deal of regulation relating to chemicals and their supporting tools, both in Europe and by and large globally, is on how 'bad' these substances are, largely based on intrinsic chemistry. This paradigm has, in turn, supported an unambitious distortion of public perception and corporate focus that sustainable development is all about 'Treading lighter on the Earth' or, in other words, being a bit less harmful tomorrow than we were yesterday. This is flawed in two principal regards. Firstly, there is no spur for paradigm-breaking innovation framed by a clear articulation of what a qualitatively different sustainable future requires and the opportunities it presents. Secondly, there is an unstated assumption of stationarity, or in other words that global carrying capacity is in a static state, when all the evidence points to an ecosystem in systematic decline and disintegration.

Whilst there are leaders in industries of all stripes choosing to make proactive commitments, there is also a substantial body of laggards waiting to be told about, or driven to do, what is required in more mundane terms. If regulation fails in its purpose of improving societal security and prospects, merely counting and only reactively responding to the most gross instances of harm or continuing to make over-simplistic judgements based on 'intrinsic properties', we are condemned to a dismal future of ever-contracting opportunity. The focus of chemical regulation then urgently needs to break out of an admittedly easier focus on intrinsic chemical properties, expanding to also address wider issues including beneficial uses that may, in reality, be context-specific to the life cycles of the products into which chemicals are incorporated.

A paper published in 2020 recognised the need for chemicals policy to drive improvements in societal resource use habits, avoiding the environmental release of toxins and recovering more substances at end-of-life of products.[8] This paper concluded that "Most instruments are reactive and fragmented", proceeding to propose a simple underpinning philosophy of "'Do no harm' regulation" including six principles outlined in Table 7.1. ("First, do no harm" is a principle applied in medicine to the actions of doctors in facing difficult decisions, deriving from the ancient Greek physician Hippocrates although it is not a part of the Hippocratic Oath.)

TABLE 7.1 Six principles for "'Do no harm' regulation" (Collins et al., 2020)

1. "Reduce and minimise releases of chemicals into the environment"
2. "Remove from use chemicals that bioaccumulate"
3. "A step change in recycling and reuse of chemicals"
4. "Use more green chemistry to manufacture greener chemicals"
5. "Commit to combined chemical and wildlife monitoring"
6. "Ensure polluters bear the full costs of prevention, mitigation and clean up"

The 'do no harm' philosophy accords with much of the 'impact question' approach applied to all product life cycle stages within the ASF approach. However, it omits the 'lost half' of sustainable development, referred to previously in this book, concerning how the use of chemicals and other societal activities can support the meeting of human, social and environmental needs in the safest and most efficient manner,[9] consistent with the bold intergenerational sense of the 'Brundtland definition'. In a suboptimal world, the balance of 'do not harm' and 'deliver needs' can sometimes co-exist in uneasy conflict. Practical examples include the use of pesticides that clearly do harm but may contribute to food security, regular treatment of biodegradable materials with biocides that are by definition harmful yet prolong useful service life and benefit delivery, transport and energy systems that are major emitters of greenhouse gases yet meet needs for mobility as well as warmth and power, and surgical procedures and many drugs that have harmful effects but can save lives or restore health. The uneasy resolution of these conflicts has direct parallels with the journey yet to be run as we aspire to a sustainable future. The same inherent conflict as well as challenges on the journey to a sustainable future is witnessed in material use by society, for which a level playing field of assessment is required to guide both choices and necessary innovations as well as, in many cases, transparent trade-offs in a world in which the goal of sustainability is substantially more than a few short steps away. Regulation is a necessary societal lever, serving both as a 'backstop' to acceptable practice as well as, ideally, a policy signal giving businesses direction and confidence for innovation and investment decisions.

A new and different paradigmatic framing of regulations is required to systemically address negative issues and positive contributions through the use of chemicals, addressing both minimisation or, ideally, elimination of harm whilst proactively optimising delivery of benefits. To implement this approach, different established models can be mobilised to implement this broader vision of chemical regulation. One such implementation pathway is found in the operation of the EU REACH regulation in Europe, specifically the requirement upon producers or importers of chemical substances to develop dossiers of evidence supporting evaluation by the European Chemicals Agency (ECHA).

If this dossier were to include the kind of systems-based analyses of risks associated with the use of substances in specific products across their life cycles, as for example outlined in the ASF approach, this evidence transparently produced by businesses would be open for public, NGO and/or regulatory scrutiny addressing the benefits of chemical use alongside associated potential threats and their mitigation measures.

The systems-based ASF tool was co-developed by TNS and VinylPlus to progress challenges associated with the use of substances, addressing wider dimensions of sustainability across whole product life cycles. It is rooted in TNS principles, offering a basis of objectivity for regulatory evaluation or, alternatively, replicable and auditable self-assessment. This approach not only provides credible design criteria for the sustainable use of additives but also identifies the weaknesses and gaps in current knowledge as well as the strengths of additives and other substances used within the full societal life cycles of product. This broader systemic framing in turn helps develop innovation roadmaps, and builds stronger communication and connections between industries, waste undertakers and recyclers, as well as regulators, regarding common life cycle sustainability objectives. A risk-based level playing field is vital for the joining up of society around common resource use challenges, be that though 'deep dive' analyses or applying ASF as a higher-level screening approach.

7.3 THE BUSINESS OF STRATEGIC REGULATION IN AN IMPERFECT WORLD

We all aspire to the attainment of sustainability for a safe and just world. However, the reality is that we live today in a far from sustainable world, with progress hampered by entrenched habits and vested interests as well as legacy assumptions, dependencies and regulations. It will take time, clear shared vision and concerted efforts to change. For this reason, a large dose of reality is required amongst regulatory architects and their governing bodies, and some of the campaigning NGOs putting pressure upon them, about feasible yet progressive steps towards longer-term goals. It is vital that choices, incentives, restrictions and actions are strategically aligned towards the goal of sustainability, but pragmatic about the fact that a single leap to attain the goal of full sustainability is vanishingly improbable in the short term.

The concern espoused by Voltaire about perfection as a potential enemy of the good has strong resonance with an inherent tension and current conflict between two flagship EU strategies relating to chemical regulation. These two strategies are principal building blocks under the European Green Deal, which comprises a range of policies across Europe and in cooperation with international partners to fight climate change.[10] The first of these strategies is the European Commission *Circular Economy Action Plan*,[11] adopted in 2020 with the intent of paving the way for a cleaner and more competitive Europe. The second is the 'clean chemistry' intent of the *Chemicals Strategy for Sustainability*, aimed towards achievement of a toxic-free, zero-pollution environment.[12] Attainment of a toxin-free environment is clearly desirable, and the increasingly circular use of resources also offers substantially lower demands on virgin resources and energy as well as waste minimisation through recovery of material value for further beneficial uses.

A problem, however, has arisen in terms of an unrealistic expectation that these two strategies, each not only laudable but necessary, should be immediately achieved. Immediate achievement of the ultimate goals of both strategies is clearly infeasible, and particularly so for long-life products – both plastics and non-plastics like – that may enter recycling streams years or decades after manufacture. These end-of-life materials may contain traces of 'legacy' substances that are now no longer used or permitted in virgin materials under 'clean chemistry' rules. Insistence on immediate achievement of 'clean chemistry' in recyclate, rather than in virgin materials, automatically kills off the equally desirable and necessary 'circular economy' aspiration. This conflict has already had a serious impact on investment by industry in recycling infrastructure, which is essential for any hope of developing an increasingly circular economy. It is doubtful that the architects of the 'clean chemistry' strategy had the intention of blocking progress towards a circular economy and, with it, abandonment of progress to date with recycling and reversion to linear resource use with its associated major waste disposal, potential pollution and resource wastage issues. However, this is the reality of an unrealistic expectation of jumping instantly to the state of perfection of 'clean chemistry', unintentionally rendered in Voltaire's terms as an enemy of the good and the sequentially improving. This problem is compounded by periodic election of new Members of the European Parliament who are propelled, by and large, with good intentions yet are naïve to some operational realities during the initiation of these strategies, and so imagining that these goals can both be attained immediately. But strategies are by dictionary definition plans of action designed to achieve long-term or overall aims, and so have to be understood and enacted as guides to progressive steps leading towards the eventual achievement of clearly articulated but aspirational goals.

Sustainable development is process, not a destination. It is a journey towards a destination of the state of sustainability – idealised today in a world still deeply mired in unsustainable norms – that can inform strategic steps feasible in the present to enable incremental and directed progress towards a clearly articulated future vision informed by backcasting. This reality is lost on those expecting instant 'cleanliness', an unintended negative consequence of which is greater reliance on virgin resources, linear resource use and disposal, rather than reuse of materials containing trace 'legacy' constituents.

A further objective reality is that not only are we contending with today's 'legacy' constituents but also, with increasingly sophisticated analytic capabilities and detection of subtle actual or likely biological effects, we will inevitably in future identify further materials of potential concern. Through absolute insistence now on 'clean chemistry' in all products, including in recyclate, we run a very real risk of setting ourselves on an unending treadmill of materials deemed acceptable today becoming future legacy constituents, condemning ourselves to unsustainable linear resource use in perpetuity. This is the case for other types of plastics and their additives, metals particularly those that are coated, substances treated with preservatives in their use phase and – as we have seen for comparative assessments on a level playing field – for many more of the materials used by society besides.

Sanity is to be found in the fact that we have to introduce a timeline towards the attainment of desirable end-goals, rather than expectation of immediate resolution between these two laudable yet currently conflicting aspirations for 'clean chemistry' and circularity. We have, in short, to take a strategic view of these two strategies. If the timeline is guided by backcasting from a clear articulation of the desired sustainable end-point – a central principle of The Natural Step approach – we can use it to navigate and inform incremental, progressive steps, recognising that our starting point is today's far from ideal norms.

Resolution of the conflict between the two European Commission strategies – *Towards a Non-toxic Environment Strategy*[13] and *Circular Economy: Implementation of the Circular Economy Action Plan*[14] – lies in recognising their explicit desirable aims but also their inherent tensions as regards the re-entry of 'problem' constituents of end-of-life products entering recycling streams.[15] Lead stabilisers in PVC products are a good example, phased out voluntarily in virgin PVC produced in Europe since before the end of 2015, but present in PVC from recovered long-life products particularly from the building and construction industries. One of the successes of VinylPlus Challenge 1/TNS Challenge 2, 'Controlled loop management', was that over 800,000 tonnes (roughly one-third of the PVC waste generated in Europe) was being recycled annually by the end of 2021 making a substantive contribution to the circular economy strategy. However, the residual lead content of this spent PVC creates an obstacle when a purist view is taken to the European Commission's *Towards a Non-Toxic Environment Strategy*, under which the slightest risk of end-of-life materials containing 'legacy' materials entering recycling routes is regarded as unacceptable. Dystopian outcomes, including greater exploitation of more virgin resource use with its 'upstream' sustainability issues, more virgin manufacturing rather than far more eco-efficient and lower-carbon recycling, and more waste generation and disposal with multiple associated 'downstream' issues, are surely not the intended model, or indeed a defensible definition, of a 'non-toxic environment'.

From a systems perspective, the fate and sustainability consequences of the estimated 150–200 million tonnes of PVC currently in beneficial, long-life uses across Europe today must be recycled into future beneficial uses to avert collateral negative consequences, including the environmental and economic costs of linear use. It is entirely possible to segregate recyclate containing a proportion of 'legacy' substances, most of which are not bioavailable as they are firmly bound within the plastic matrix, and put it to safe uses in recycled PVC products for which a proportion of lead is not problematic. Further research is occurring to enhance the detection of legacy constituents in recovered PVC. Were this more rational path followed, PVC material entering recycling streams would be of increasing purity with respect to legacy materials as a greater proportion of end-of-life PVC produced under a 'clean chemistry' strategy enters 'circular economy' recycling loops. There is also the option of blending virgin and recovered sources to further reduce legacy content for specific applications. This is already the approach taken as a norm in treatment of water resources reticulated in public supply networks, widely and safely blended from more and less enriched sources to meet safety standards. This 'glide path' strategy towards incrementally cleaner recycled chemical product represents a strategic and yet pragmatic step towards a vision of sustainability, rather than a wholly unrealistic expectation of immediate achievement. The 'glide path' approach thereby resolves the current conflict – perceived rather than actual though actualised by unrealistic expectations – between strategic European Commission chemical strategies simply by recognising the objective reality that full realisation of end-goals will necessarily take time from the starting point of the world in which we actually live today.[16] That, as has been said, is what a strategy is meant to achieve.

A further major benefit arising from unblocking unnecessarily obstacles to stepwise progress towards circularity, imposed by the naïve belief that sustainable end-goals can be attained by an immediate step into 'clean chemistry' from today's realities, is that recycling of PVC has huge benefits in terms of carbon intensity relative to continued dependence on virgin product and the consequent generation of wastes and wastage of spent resources. Progress towards climate stability is another environmental benefit of growing political importance that favours the strategic contributions achieved by increasing circularity. These important aspirations must not be slaughtered on the altar of expectation of immediate attainment of 'clean chemistry'.

Regulatory acknowledgement of these realities would also, importantly, offer reassurance to businesses investing in the recycling infrastructure essential for progress towards the achievement of an increasingly circular economy, as well as contributing to many linked sustainable development challenges. The key questions for policy-makers and influencers are, simply, whether sustainability is the intended end-goal and, if so, how can it be achieved on a progressive journey rather than be thwarted by an inflexible expectation of a leap towards chemical purity – unjustified both scientifically and economically – with associated wastage of value and material supporting further beneficial societal uses.

7.4 THE BUSINESS OF REGULATION IN A GLOBALISED WORLD

An issue of particular concern relates to products bought online, or conceivably imported in bulk, that cross borders into regions aspiring to more sustainable practices. Substances and products, both plastic and non-plastic, may derive from geographical regions where there is little or far-reduced scrutiny of chemistry, pollution and ethics in supply chains and manufacture.

In a globalised world, this loophole must be plugged by effective regulation if the sustainability benefits of responsible PVC production and use are to be realised. Otherwise, responsible business operators may simply be undercut by cheap imports from regions where such practices are not observed. A level playing field needs to apply, ultimately, on a pan-global basis or, at minimum, at the borders of international trade. This is not a matter of protectionism. It reflects a deeper commitment to sustainable development, under which societal and environmental wellbeing are not simply overlooked by a narrow focus on lowest financial cost. Creation of quality assurance rules relating to the sustainability footprint of internationally traded goods is a form of regulation with purpose and vision. It has parallels, for example, with CITES (the Convention on International Trade in Endangered Species), adopted in 1963 as an intergovernmental agreement to ensure that international trade in specimens of, or products derived from, wild animals and plants does not threaten the survival of scheduled species.[17] A range of other intergovernmental agreements and protocols relate to environmental, ethical and other aspects of sustainable development. A strategically informed level playing field for trading of chemicals, including those embedded in products, needs to be enacted urgently. This needs to occur not just in national and geo-regional policies but, for example, throughout the global rules of trade between nations under the World Trade Organization (WTO).[18]

7.5 A GLOBAL CHALLENGE

The use of chemicals and materials is a theme affecting the whole world, contributory to both its sustainability threats and solutions. There is an urgent need for a level playing field, transparently founded on scientific principles to address the wider dimensions of sustainable development across whole material life cycles. Without it, we will continue to innovate what is 'best' without counting the upstream, in use and downstream costs of the substances we deploy to meet our wants and needs.

The history of regulation and voluntary commitments, as well as consumer choices and corporate actions, has taken us some way on a journey of improvement since the dawning days of awareness of the unintended consequences of society's profligate resource use habits. This has happened in waves, variously from the time of the first Alkali Act in the UK through to the substantial boost in environmental awareness of the 1970s, and onwards to stated commitments at international, national, regional and more decentralised scales to the principles of sustainable development.

There is an urgent need for a level playing field that is applicable and that is also then practically applied globally. Internationalisation of consistent, scientifically founded and practical methods to assess the sustainability of material use across societal life cycles, be that regulatory or voluntary, is far more important than just thinking within borders, as the scale of burgeoning human pressures on the world is truly global in scale and threat. These dangers range from international emissions of climate-active gases, waste and pollution of various types with transboundary consequences, depletion of natural resources and biodiversity, to ethical issues along supply chains from raw material extraction through manufacturing and onwards to exposure in the use or post-use of products.

The achievement of consensus at an international scale about assessment of the sustainable use of all materials would make a significant contribution towards concerted action to achieve sustainability.

7.6 THE BUSINESS OF LEADERSHIP

One of the lessons that I learned from my 22 years in the public sector, including in regulatory organisations and national government as well as serving as an international government advisor, is that governments never, or at least only rarely, lead. NGO pressure focusing public concern, often amplified by media, has a long history of raising issues of concern. Research can highlight issues, often in concert, as a knowledge translator, with citizen-based campaigning bodies. Industry can innovate and petition for regulatory change to set a performance baseline, dissuading unfair competition from less responsible producers. Intergovernmental bodies can forge agreements requiring ratification and response by national governments. However, innovation to meet new societal expectations, strategic priorities and standards is predominantly the purview of business.

Business is often justifiably accused of resistance to change to protect short-term profit-taking, as for example the powerful influence of some sectors of industry lobbying to dilute or derail stated political commitment to tackle climate change. But it is also true that responsible and far-sighted businesses are, and have been, pioneers of innovation and systemic change, as for example recent step-changes in renewable energy generation and battery technologies that are now undercutting the price-performance of established, more polluting energy generation technologies. One of the most impressive recent examples is the pace of innovation, approval and roll-out of Covid-19 vaccines, showing what is possible when societal sectors and nations work together to address urgent and clearly articulated needs.

The role of activists and consultants like me may be to offer strategic insight and to serve as a critical friend to businesses, but I am under no illusion that it is business that will invest the financial and human resources necessary to deliver innovative products and services helping society make progress towards sustainability goals. It is my sincere hope that I can help leading businesses become more profitable through better serving societal needs safely and efficiently. The pioneers can thereby reshape the operating environment within which wider foresighted businesses can or will need to adapt, and that leaves other enterprises with lesser concern for such matters to founder in markets inevitably and increasing reshaped by impinging sustainability pressures.

NOTES

1 Everard, M. (2000). Aquatic ecology, economy and society: the place of aquatic ecology in the sustainability agenda. *Freshwater Forum*, 13, pp. 31–46.
2 Porritt. (2007). *Capitalism as if the World Matters*. Routledge.
3 Parliament of the United Kingdom. (1863). *An Act for the More Effectual Condensation of Muriatic Acid Gas in Alkali Works*. Parliament of the United Kingdom.
4 Wolf, M. (2023). *The Crisis of Democratic Capitalism*. Allen Lane.
5 Fukuyama, F. (1992). *The End of History and the Last Man*. Free Press.

6 Mason, R. (2013). David Cameron at centre of 'get rid of all the green crap' storm. *The Guardian*, 21 November 2013. [Online.] https://www.theguardian.com/environment/2013/nov/21/david-cameron-green-crap-comments-storm, accessed 14 April 2023.

7 itvNEWS. (2013). 'Giving up on waterways': Ministers scrap anti-pollution housebuilding rules. *itvNews*, 29 August 2023. [Online.] https://www.itv.com/news/2023-08-29/giving-up-on-waterways-ministers-scrap-anti-pollution-housebuilding-rules, accessed 30 August 2023.

8 Collins, C., Depledge, M., Fraser, R., Johnson, A., Hutchison, G., Matthiessen, P., Murphy, R., Owens, S. and Sumpter, J. (2020). Key actions for a sustainable chemicals policy. *Environment International*, 137, p. 105463. DOI: https://doi.org/10.1016/j.envint.2020.105463.

9 Everard, M. and Longhurst, J.W.S. (2018). Reasserting the primacy of human needs to reclaim the 'lost half' of sustainable development. *Science of the Total Environment*, 621, pp. 1243–1254. DOI: https://doi.org/10.1016/j.scitotenv.2017.10.104.

10 European Commission. (2020). *European Green Deal*. European Commission. [Online.] https://climate.ec.europa.eu/eu-action/european-green-deal_en, accessed 13 March 2023.

11 European Commission. (2020). *Circular Economy Action Plan*. European Commission. [Online.] https://environment.ec.europa.eu/strategy/circular-economy-action-plan_en, accessed 13 March 2023.

12 European Commission. (2020). *Communication from the Commission to the European Parliament, the Council, the European Economic and Social Committee and the Committee of the Regions: Chemicals Strategy for Sustainability towards a Toxic-Free Environment*. European Commission, Brussels, 14.10.2020 COM(2020) 667 final. [Online.] https://circabc.europa.eu/ui/group/8ee3c69a-bccb-4f22-89ca-277e35de7c63/library/dd074f3d-0cc9-4df2-b056-dabcacfc99b6/details?download=true, accessed 13 March 2023.

13 European Commission. (2017). *Towards a Non-toxic Environment Strategy*. European Commission, Brussels. 2017. http://ec.europa.eu/environment/chemicals/non-toxic/index_en.htm.

14 European Commission. (2018). *Circular Economy: Implementation of the Circular Economy Action Plan*. European Commission, Brussels. 2018. http://ec.europa.eu/environment/circular-economy/index_en.htm.

15 Smith, N.C. and Jarisch, D. (2016). INEOS ChlorVinyls: a positive vision for PVC (A). *Managing Sustainable Business*, 2016, pp. 73–106.

16 Everard, M. (2020). A lead on recycling PVC. *Materials World*, May 2020, p. 47.

17 CITES. (2023). *What Is CITES?* The Convention on International Trade in Endangered Species (CITES). [Online.] https://cites.org/eng/disc/what.php, accessed 14 April 2023.

18 WTO. (n.d.). *The WTO*. World Trade Organization (WTO). [Online.] https://www.wto.org/english/thewto_e/thewto_e.htm, accessed 17 March 2023.

Epilogue

8

Although appearing within its subtitle, this book is not really about PVC. It is informed by a great deal of the issues relating to this controversial material, which drove those engaged in the PVC value chain to recognise that sustainable development set a strategic context for necessary action and innovation. But innovative thinking, commitments and practices associated with this plastic over the past quarter-century have informed a far broader context, elevating thinking to a different level addressing the need for, and the characteristics of, a level playing field against which to assess how the many substances that society uses can be selected, innovated and used wisely to best meet its various needs in the most efficient and safest manner. The underlying theme of this book is, in essence, 'material blind'; it is concerned with a more deeply connected paradigm of thinking about what is optimally sustainable.

Underneath it all, I am just a fish-hugger. A great deal of my life, research and development support has been about water in its widest sense, from aquatic ecology to its role as a vital yet threatened natural resource underpinning the meeting of needs in the developing and the developed worlds alike. But, as a systems scientist who committed early to promotion of sustainable development, I have had the privilege of working with people and organisations to pursue that goal across multiple societal sectors and in every continent on Earth (except the poles). Whilst I am not a chemist by background, I have been working with teams of brilliant chemists in tackling topics including those covered by this book.

Over the years, I have faced 'friendly fire' from colleagues in the 'green movement' who felt that the PVC industry was on the ropes and ready for a knockout blow. As you may recall from earlier in this book, the UK PVC industry felt that this was a potential outcome when I was invited to that confidential meeting in 1999 to help it recognise and address its sustainable development challenges. My work has been part of a collective effort by many others to embed real, science-based principles, commitments and objectively audited targets across the PVC sector both in Europe and with growing influence on the global stage. For this, I have been accused of being pro-PVC.

DOI: 10.1201/9781003453949-8

Let me be clear on that point up: I am not pro-PVC. I am simply anti-stupidity!

I have seen too many times, in my life-long involvement promoting what we now call sustainable development (but had not yet called it that when I started), that a 'bad' material or practice pilloried and banned in the public eye is often substituted with an alternative that has not been assessed on a level playing field of principles. Time and again, businesses and other institutions simply reinvest in tomorrow's unassessed potential problem material, process or issue, due to lack of vision of what truly defines a sustainable outcome. I have worked with the PVC sector to avert this trap and, as a critical friend, I have also regularly reminded players in the sector that there are real material-specific issues that need to be addressed. If I am guilty of driving progress in the sector, I can only apologise to my critics!

However, the prize of this quarter-century of engagement in the PVC sector is bigger than that, and is beginning to be realised as more material sectors come to recognise that media and NGO attention lavished elsewhere, on another focal material than their own, will not ultimately exempt them from being forced to confront and address their own sustainability challenges. As we saw when the TNS Sustainability Challenges for PVC were applied to proposed substitute materials, as well as a rapid application of the ASF (Additive Sustainability Footprint) tool to a range of other materials, all in reality share a common set of sustainability challenges that vary by the details of their supply chains, maintenance requirements, potential for recyclability, some aspects of unique chemistry and other life cycle facets. If I am guilty of influencing the way in which we re-evaluate the implications of the use of all materials on a common level playing field, then sorry again: I stand guilty as charged!

In conclusion, we live at a pivotal time in human history when, faced with daunting interlinked, potentially existential challenges, we can no longer afford facile material-against-material squabbles. To grapple with the daunting sustainability challenges facing global society, we need to be mature enough to evaluate, on a transparent basis, how best to service people's needs in the safest and most efficient manner. Whether we are in the chemical and materials industries or other sectors of business, or in regulatory or government bodies seeking how most effectively to propel society forwards to meet its needs in equitable and optimal ways, we need this deeper insight and commitment to think beyond today's dangerously simplistic norms and assumptions, and to reach for what is truly sustainable.

That is the great prize and the deeper purpose of this book. May we all find the wisdom and courage to work together to achieve that goal, to safeguard a future of greater security and opportunity for all.

Index

Note: **Bold** page numbers refer to tables and *italic* page numbers refer to figures.

A, B, C, D approach 26
accreditation 3, 43, 64–67
acetylene carbine process 34
acetylsalicylic acid (ASA) 8
'acid-free fumes' 52
activism 41, 43
additives 1–3, 12, 21, 33–37, 39–41, 45, 48,
        49, 53, 64, 66–70, **71–74**, *75*,
        80–82, 90, 92–96, 100, **102–106**,
        118, 120
Additive Sustainability Footprint (ASF) 3,
        67–70, *68,* **71–74**, *75,* 82, 96–108,
        117, 118, 127
adiabatic cracker **50**
African elephant(s) 7
age of plastics 2, 11
agricultural revolutions 10
Akdeniz Chemson **77**
alcohol 34, 36
Alkali Act 113, 123
alternatives to PVC 41, 89, 92, 95
Altro 82
aluminium 20, 40, 54
America 20, 23, 41, 86, 90, 95, 108, 113
Ancient Greek 10, 116
animal welfare 52
antibiotic(s) 8, 16
antioxidant(s) 35
anti-pollution 114
antistatic agent(s) 36
anti-stupidity 127
arum lily 14–15, *15*
Arum maculatum see arum lily
asbestos, neonicotinoid pesticides 22
ASEAN *see* Association of Southeast Asian
        Nations (ASEAN)
ASEAN Vinyl Council (AVC) 84
Asociacion Argentina del PVC 84
aspirin (acetylsalicylic acid, ASA) 8
Assessment of the sustainable use of
        chemicals on a level playing field
        69, 97, 100

Association of Southeast Asian Nations
        (ASEAN) 84, 86
ASUA 81
'Attenborough Effect' 39
Attenborough, Sir David 39
audit 30, 48, 91
audited 44, 59, 60, 66, 67, 92, 115, 126
auditor 65
Australia 84, 86
autotroph(s) 5, 6
aviation 11, 28, 37
Ayurvedic medicine 17

backcasting 24–27, 46, 70, 80, 83, 120
badger(s) 6
Baekeland, Leo Hendrik 11
Baerlocher **77**
Bakelite 11
battery technologies 124
beaver(s) 7
behaviour(s) 20, 21, 28, 47, 52, 54
bioaccumlative 38
bioaccumulation 38
biocidal 24
biocidal preservative(s) 21, 101
biocide(s) 21, 99, 117
biology 5, 6, 8
biomineralisation 26
biosphere 25, 83
BIOVYN™ 75
bird biomass 18
birth defect(s) 38, 80
blood bags 12, 37, 80
blowing agent(s) 36
blow moulding 36
blown film production 36
brominated flame retardants 81, 94, 101
Bronze Age 9, 10
Brundtland definition 29, 62, 117
'building block' molecules 26
building product(s) 65, 99
burning, open 38

business 2–4, 20–22, 24, 25, 28–30, 40, 46,
  80, 83, 89, 108, 112–116, 127
  global challenge 123
  globalised world 122
  leadership 123–124
  regulation 116–118
  strategies 118–122

cable(s) 52, 92, 94, 100
calcium carbonate **50**
calcium/organic 81
calcium/zinc 81
calendering 36
Cameron, David 114
Canadian Chemical Producers' Association
  (CCPA) 47
cancer(s) 17, 38
cannulas 12, 37
capitalist model 3, 112, 113, 115
carbohydrate(s) 11
carbon 10, 26, 29, 30, 37, 46, 81, 92, 94, 121
carbon dioxide (CO_2) 5, 30, 81, 82
carbon dioxide equivalent (CO_2e) 30
carbon footprint 29, 30, 40, 75
carbon handprint 29, 30
carbon neutrality 30, 75, 80, 93
carcinogenic, mutagenic and reprotoxic
  properties (CMR) 28
Carson, Rachel 23, 37
cast iron 44
catheters 12, 37, 80
caustic soda 34
CCPA *see* Canadian Chemical Producers'
  Association (CCPA)
CDs 11
cellophane 11
cellulose 11
CEN EN 13501-1EN standard 94
ceramics 8, 12
CFC *see* chlorofluorocarbon (CFC)
chalk 36
chemical assessment approaches 97,
  98–99, 100
Chemicals strategy for sustainability 119
Chemical Weapons Convention 37
chemistry 1, 2, 5–32, 38, 53, 67, 86, 90, 97,
  116, 122, 127
Chemistry Industry Association of Canada 47
Chernobyl 6
China 10, 34
chlorinated paraffins 80–81
chlorinated polyvinyl (CPVC) 94

chlorine free radicals 38
chlorine gas 37
chlorofluorocarbon (CFC) 25, 38
chlorophyll 38
circular economy 39, 40, 75, 90,
  119–122
Circular Economy Action Plan 119
*Circular Economy: Implementation of the
  Circular Economy Action
  Plan* 120
Circular Economy Law 47
Circular Flooring Project 82
Circular Plastics Alliance 81
circular use 28, 99, 119
CITES *see* Convention on International Trade
  in Endangered Species (CITES)
'clean chemistry' 119–121
climate-active 22, 29, 30, 123
climate change 18, 20, 22, 23, 46, 52, 55,
  119, 124
climate-forcing 25
Club of Rome 23
cobalt 100
coffee 16, 17
colour coding 69–70, 97, **98–99**
compatibiliser(s) 36
composite(s) 12, 93
Computerised Tomography (CT) 12
concrete 8
consensus 25, 29, 61, 123
conservation 25, 52, 83, 114
Conservative political party 114
construction 8, 11, 28, 37, 43, 44, 54, 90, 92,
  99, 120
Construction Product Regulation (CPR) 48
controlled loop 40, 70, 81, 101
  management 120
Convention on International Trade in
  Endangered Species (CITES) 122
copper 9, 41, 95
Copper Age 9
coral(s) 7
Corporate Societal Responsibility (CSR) 115
corrode 93
corrosion 93
Covid-19 18, 86, 124
CPVC *see* chlorinated polyvinyl (CPVC)
Cradle to Cradle **98**
CreaSolv® 82
crocodile(s) 6
cuckoo-pint *see* arum lily
customer demand(s) 108

DDT 23
decarbonisation 46, 80
decision-making 13, 22, 25, 49, 55, 89
degradation 2, 18, 19, 35, 93
DEHP (Di (2-ethylhexyl) phthalate) 39, 80, 82
DEKRA GmbH **76**
Denmark 41, 44
de-risk (de-risking) 4, 116
'dirty secrets' 115
disposal 19, 38, 43, 45, 47, 54, 82, 85, 90, 95,
     119–121
'do no harm' regulation 17, 117, **117**
drainage pipes 41
drugs 8, 11, 12, 17, 117
ductile iron 93–94
DuPont company 108
durability 12, 24, 29, 35, 39, 42, 48, 53, 54,
     95, 107
durable material(s) 20, 21, 40
DVDs 11

Earth's crust 93
ECHA Plastic Additives Initiative **98**
Ecodesign Directive 47
*Ecodesign for Sustainable Products
     Regulation* 47
eco-efficiency/eco-efficient 36, 45, 107, 121
*Eco-efficiency Code of Practice for the
     Manufacture of PVC* 44, 45, 48
ecosystem degradation 18, 19
ecosystem engineers 7
The Ecosystems Revolution 9
ecotoxicity 28
Ecovadis **99**
EcoVin® **50**
ECVM Charter 58
Egypt (Egyptian) 9, 10
elastic 34
electronics 11, 12
EMAS-compliant 43
emission(s) 19, 22, 29, 30, 36, 40, 41, 45, 46,
     58, 59, 81, 82, 86, 90, 94, 123
emulsion PVC (E-PVC) 58
'end of history' 114
End of Life Vehicles Directive 40
energy 5–7, 11, 17, 20, 21, 24–26, 29, 36, 37,
     40, 81, 83, 93, 94, 101, 107, 114,
     117, 119, 124
Environment Agency 43, 80
*Environmental Charter for UK PVC
     Manufacturers* 44, 45, 48
environmental impacts 38, 44, 84, 92

Environmental Issues of PVC 40, 90
Environmental Management Systems 43
Environmental Product Declaration (EPD) 28
Environmental Protection Agency (EPA) 23,
     41, 90, 107
'environment movement' 2, 22, 52
E-PVC *see* emulsion PVC (E-PVC)
ethics 19, 86, 122
ethylene dichloride (EDC) 34, *34*
ethylene process 34, *34*
EU Member States 40, 43–44
Europe 2, 10, 14, 20, 25, 38, 39, 41–43,
     52–54, 58, 80, 82, 85, 86, 90, 107,
     116, 117, 119–121, 126
European Chemicals Agency (ECHA) 117
European Commission 40, 43, 44, 47, 48, 90,
     119–121
European Council for Plasticisers and
     Intermediates (ECPI) 58
European Council of Vinyl Manufacturers
     (ECVM) 58
European Green Deal 46, 119
European Industrial Revolution 10, 112
European Plastics Converters (EuPC) 58
European Resilient Flooring Manufacturers'
     Institute (ERFMI) 82
European Stabiliser Producers Association
     (ESPA) 58
European Union 36, 82, 107, 118
eutrophication 28
EVC 43, 45
Everard, Mark 49, *52*
extended producer responsibility 47
extrusion 36, 82

faeces 15, 16
farming 19, 62, 95
felling 7, 19
Fertile Crescent 10
filler(s) 36, 80
fire-quenching 94
First World War 11, 37
fisheries 46, 52, 115
fish-hugger concept 126
flame retardant(s) 21, 36, 53, 101
flammable/flammability 52, 94, 101
flexible 29, 35, 36, 93
fontanellar gun 6
food web(s) 6
forest-based products 92–93
Forestry Stewardship Council (FSC) 45–46,
     64, 65, 93, 115

'forever chemicals' 22, 108
formaldehyde 11
Forum for the Future 43
fox(es) 6
Framework for Strategic Sustainable
        Development (FSSD) 27, 45, 92
freedom of operation 20
free-market capitalism 113
French Enlightenment 55
French Environmental Code (FEC) 47
Friedman, Milton 113
'friendly fire' 126
fugitive emission(s) 40, 45
Fukuyama, Francis 114
fungi 6, 7, 16, 17
'The Funnel' 21, 26, 45, 108
future-proofing 49

gamma radiation 6
geomembrane(s) 11, 37
germanium 12
glass 12, 21
'glide path' 121
globalised world 122
global recession 114
global resource extraction 18
Global Vinyl Council (GVC) 84–85
Gothenburg Protocol 93
'grandparenting' 23, 108
Great Acceleration 19
'Greed is good' 113
Green Chemistry and Commerce Council
        (GC3) Retailer Database **98**
greenhouse gas(es) 30, 46, 117
'green movement' 126
Green Paper 40, 90
Greenpeace 41–43, 89, 92
GreenScreen List Translator™ **98**
Greensuite® **98**
GreenWERKS **98**
GRI 301: Materials **99**
guttering 41

halides 38
hazard 23, 29, 38, 67, 68, 97
HCFC *see* hydrocholorofluorocarbon (HCFC)
health and safety 85, 113
healthcare 11, 12, 37, 62, 90
'heat map' 69, *75*
heavy metals 26, 45, 80
heterotrophs 6
Hippocrates 116

Homo erectus 9
homogeneous PVC flooring **77**
human demand 3, 20, 89
human exploitation 2, 14–17, 21
human health 2, 18, 23, 40, 42, 97, 99, 113
    and environment 12–13, 28–30
    exploitation of natural chemistry 14–17
    social abuse 22–28
    use and abuse of chemicals 9–12, 17–22
human rights 113
human skills 8
hydrocholorofluorocarbon (HCFC) 25
hydrogen 34, 36
hydrogen chloride 34, 38, 67
Hydro Polymers 43, 45, 49, **50–51,** 59

ICCA Plastics Additives Working Group 85
ideology 113
IKA *77*
impact modifier(s) 12, 36
'impact questions' 69, 96, 117
implants 12, 37
India 17, 62, 83, 86
Indian Vinyl Council 83
Indonesia 16
Industrial Revolution 19, 115
INEOS Inovyn 43, 75
inequities 19
inert 7, 12, 23, 24, 36, 38, 40, 68, 80, 99
injection moulding 36
innovation 1–4, 8–13, 17, 20–22, 28–30, 33,
        34, 37, 55, 59, 62, 68–70, **71–74,**
        75, 80–83, 89, 101, 107–108,
        116–118, 124, 126
Institute of Directors 45, 64
Instituto Brasileiro 84
insulation 11, 12, 21, 24, 37, 38, 94
intergovernmental agreement 122
International Council of Chemical
        Associations (ICCA) 85
The International Council of Chemical
        Associations (ICCA) 85
International Council on Mining & Metals 94
internationalisation 123
international trade 122
intrinsic properties 23, 62, 97, 116
investment 3, 4, 22, 25, 33, 40, 48, 49, 53, 69,
        82, 89, 90, 108, 113, 114, 117, 119
ionising radiation 38
Iron Age 10
ISO9001 43
ISO14001 43

ISO14040:2006 life cycle stages 68, *68*
ISO 14040 standard 96
Italian Official Journal 66

kelp bed(s) 7
Klatte, Friedrich 33
'knee-jerk' 22, 25, 53
knowledge translator 124

labour 21, 52, 113
Law on the Circular economy 47
lead 4, 20–22, 25, 27–29, 39, 40, 45, 47, 53,
        80, 82, 85, 120, 121, 124
leadership 49
    business 123–124
    role 39
leather 12
leather cloth 36
'legacy' additives 39, 82
legacy constituents 120, 121
legislation 40, 47, 48, 55, 113
level playing field 2–4, 22, 24, 41, 42, 55, 63,
        85, 89–91, 95–107, 108, 117, 118,
        120, 122, 123, 126, 127
    'traffic lights' colour coding 69–70, 97,
        98–99
liabilities 47, 89, 108
life cycle(s) 3, 13, 20, 21, 24, 39–41, 43, 45,
        46, 48, 53–55, 64, 67, 68, *68,*
        69, 70, **71–74,** 75, 80–83, 89, 90,
        94, 95, 97, 100, 101, 107–108,
        116–118, 123, 127
    positive and negative implications 101,
        102–106
    risk approach 108
life cycle assessment (LCA) 28, 43, 44, 68,
        69, 95–96
*Life Cycle Assessment of PVC and of
        principal competing materials*
        43, 90
Limits to Growth 23
linear resource use 19, 20, 119, 120
linoleum 44, 95
livestock 10, 18, 22
lobby 22
lobbying 124
lords-and-ladies *see* arum lily
'lost half' of sustainable development 29, 53,
        61, 97, 100, 117
lubricant(s) 12, 35

Macbeth 17
Magnetic Resonance Imaging (MRI) 12
maintenance 13, 19, 21, 24, 29, 39, 44, 53, 54,
        69, 90, 92, 95, 127
mammalian biomass 18
marine litter 39, 40, 94
Marine Stewardship Council (MSC) 46, 64,
        65, 115
markets 3, 20, 21, 28, 29, 44, 47, 52, 53, 62,
        65, 84, 92, 93, 112–114, 124
mass extinction 18
'material blind' 4, 126
material choice 2, 4, 22, 108
material flow cost accounting (MFCA) **99**
matter conservation 25, 83
medical 1, 8, 12, 36, 37, 44, 54, 80, 99
    application(s) 36, 37, 54
medical imaging technologies 12
medicine(s) 8, 11, 17, 116
Medicines & Healthcare Product Regulatory
        Agency 80
melanin 6
Members of the European Parliament 119
Mesopotamia (Mesopotamian) 9, 10
metal-based stabiliser 100
microplastics 40
'middle class' lifestyles 18
Minamata Convention 93
mined cobalt 100
mining 19
mitigation 3, 70, **71–74,** 101, **102–106,** 115,
        118
monetarism 113
monetarist 114
Montreal Protocol 25, 38
mouldable 34
multinational agreements 114
multinational corporation(s) 113
municipalities 28
muriatic acid 113

National Centre for Business and Ecology
        (NCBE) 42–44, 90
national institutions 84
natural capacity 20
natural polymers 11
nematode(s) 8
neoliberalisation 113
nerve gas(es) 37
Netherlands, The 36

NGO(s) 2, 4, 25, 42, 43, 48, 49, 52, 53, 89, 90, 118, 124, 127
  sectors 37–41
Nixon, Richard (President) 23
Nobel Prize 113
non-PVC plastic 94
nonsteroidal anti-inflammatory drug (NSAID) 8
non-toxic environment 121
Norsk Hydro 49
North America 83
Norwegian Environment Agency 80
nutrient substance(s) 26

OECD *see* Organisation for Economic Co-operation and Development (OECD)
OECD Substitution and Alternatives Assessment Toolbox (SAAT) **98**
one-pack 81
Organisation for Economic Co-operation and Development (OECD) 20
organochlorine substance(s) 8, 37, 38
ozone depletion 25, 28, 52
ozone layer 25, 38

packaging 11, 37, 43, 81, 92
paint 21, 39, 92
paradigmatic change 3, 29
Paris Agreement 20
PCBs *see* polychlorinated biphenyls (PCBs)
peer-reviewed paper 91, 97, 101, 108
perfluorooctanoic acid (PFOA) (C8) 108
The Perils of PVC Plastic Pipes 95
permeability 18, 36
persistence 1, 24, 86
persistent organic compounds 48, 95
persistent substance(s) 23, 24, 29, 93, 108
pesticides 8, 17, 22–24, 29, 38, 39, 52, 107, 117
petrochemical 22
pharmaceutical(s) 11, 107
phenol 11
pheromone 16
photosynthesis 5, 6
physics 5, 6, 8
pigment(s) 6, 12, 35
pipe(s) 11, 24, 36, 37, 41, 44, 62, *63,* 92–95
plasticisers 12, 35, 39, 40, 44, 58, 80, 82
plastic pollution 40
PlasticsEurope 95
plastic waste 1, 20, 39, 40, 54, 85

plumbing 11, 37
poison(s) 17
'political football' 41
pollen 15, 16
pollination 16
pollution 19, 20, 40, 86, 101, 113, 119, 122, 123
polychlorinated biphenyls (PCBs) 38
polyethylene 37, 94
Polyflor 82
polymer(s) 2, 8, 11, 12, 33–38, 40, 44, 54, 75, 80, 82, 93, 94
Polymer Chemie *77*
polyolefin(s) 36, 37, 54, 94–96, 101
Polyolefin Circular Economy Platform (PCEP) 95
polyoxybenzylmethylenglycolanhydride 11
polypropylene 37, 94
polyvinyl chloride (PVC) 1, 2, 4, 8, 28, 33–88, *34,* 49, *52,* 89, 90–97, 99, 100, 101, **102–106**, 101, 115, 120–122, 126, 127
  Europe and UK 41–43
  global regions 83–86
  industry 115, 126
  NGO sectors 37–41
  optional additives 35–36
  polymer 35
  prejudgements 52–55
  sector 127
  sustainability challenges 43–52, 91–96
  TNS Sustainability Challenges 60, **61**
Porritt, Jonathon 43, 45
post-consumer 82, 85
post-industrial **71–72, 102, 105**
post-Second World War 114
power supply 11
pre-consumer 85
preservative(s) 17, 21, 92, 93, 99, 101, 120
pressure group(s) 41, 42, 48, 53, 59
pressure treatment 21, 92
price-performance 124
Prime Minister 113, 114
printing 11
processing aids 35, 68
procurement 4, 66
Product Environmental Footprint (PEF) 28, 65
product life cycle(s) 3, 48, 53, 55, 68, 70, 97, 100, 101, **102–106,** 101, 107, 108, 117, 118
Product Stewardship 47, 84

profit 47, 112–115
profitable 2, 4, 21, 108, 114, 116, 124
'progress questions' 69, 96
propellant(s) 12, 38
pro-PVC 126, 127
protoplanet 26
public health 22, 25, 47
publicity 108
PVC *see* polyvinyl chloride (PVC)
PVC: An Evaluation Using The Natural Step Framework 46
PVC Coordination Group 43–45, 48
PVC for Tomorrow 49
PVCMed Alliance 54
PVC: Reaching for Sustainability 49, 52
PVC Retailer Working Group 42, 43
PVC Stakeholder Forum for Sustainability 48, 49

quality assurance 83, 122
Quality Management Systems 43

rabbit(s) 6
radiotrophs 6
Rafflesia arnoldii 16
raw material 1, 19, 24, 44, 64, 90, 95, 107, 123
REACH *see* Registration, Evaluation, Authorisation and Restriction of Chemicals (REACH)
Reagan, Ronald 113
Reagens **77**
recession 114
Recofloor™ 82
RecoTrace™ 81
recovery 1, 20, 39, 40, 53–55, 69, 90, 92–95, 107, 114, 119
Recovinyl® 81
recyclability 21, 24, 29, 37, 48, 53, 54, 75, 81, 92, 94, 127
recyclate(s) 81, 119–121
recycling 1, 20, 24, 28, 29, 39, 40, 43, 44, 47, 53–55, 59, 69, 70, 81, 82, 85, 90, 92–95, 101, 107, 119–122
    chemical 44
    mechanical 44
refrigerant(s) 38
Registration, Evaluation, Authorisation and Restriction of Chemicals (REACH) 23, 28, 69, 80, 82, 107, 117
Regnault, Henri Victor 33

regulation 3, 4, 23, 29, 41, 46, 47, 59, 80, 91, 94, 97, 107–108, 112–125
    business 116–118
regulatory attention 1, 115
REMADYL 82
renewable 75, 101, 124
renewable energy generation 124
renewable energy technologies 11, 37
Repurposing business around the meeting of human needs 62
resource(s) 1, 3, 7, 13, 18–21, 49, 75, 82, 83, 86, 89, 95, 100, 107, 112, 113, 115, 116, 118–121, 123, 124, 126
    depletion 19, 21, 100
Responsible Care programme 47, 85
restriction(s) 38, 108, 118
retrospective regulation 113
revolution(s) 9–11, 112
rhizosphere 7, 8
rigid 35, 36, 82
'Rio Conference' 114
risk(s) 3, 19, 22–24, 28, 38, 40, 43, 47, 48, 53, 59, 67, 68, 80, 90, 95, 97, 99, 107, 108, 114, 118, 120
    analysis of 53
Robèrt, Dr Karl-Henrik 25
rotation moulding 36
Roundtable on Sustainable Biomaterials (RSB) 75
rubber 11, 12, 34

Salix genus 8
Saving Our Skins 41
scent 12, 16
Scheme Management Contribution 65
SciveraLENS® **98**
semiconductor 12
service life 20, 21, 24, 37, 39, 53, 54, 69, 70, 92–93, 99, 101, 117
Shakespeare, W. 17
Silent Spring 23, 37–38
silicon 12
single-use plastic 29, 40
slavery 21
SLCA approach 96, 97, 100
snapshot report, ASF 69
social abuse, human health 22–28
societal value 93, 107
Southern Africa 84
Southern African Vinyls Association (SAVA) 84
Soviet Union 114

spadix 14, 15, *15*
spigin 6
spraint(s) 16
S-PVC (suspension PVC) 58
stabilisation 35, 39, 81, 85
stabilisers 12, 35, 39, 44, 69, 70, *71–74,*
　　80–82, 85, 101, 120
static electrical charge 36
status quo 22
steel 8, 10, 54
step-changes 124
stepping stones 25
sticklebacks 6
'Stockholm Conference' (the UN Conference
　　on the Human Environment) 23
stone 9, 10, 12, 25
Stone Age 9, 12
Stone-Bronze-Iron system 9
strategy 25, 70, 114, 118–121
sub-Saharan region 84
substitution 23, 42, 53, 89
supply chain(s) 19, 24, 48, 49, 53, 55,
　　69, 75, 86, 95, 100, 115, 122,
　　123, 127
surfactant(s) 54
sustainability 1–4, 13, 18–21, 24, 25, 27,
　　37, 41–43, 45, 46, 48, 49, *52,* 53,
　　55, 58–97, 100, **102–106**, 101,
　　107–108, 112–125, 127
　challenges for PVC 43–52, 91–96
　positive and negative implications 101,
　　102–106
sustainability assessment 1–3, 24, 28, 90,
　　96, 107
sustainability footprint(s) 21, 24, 29, 48, 54,
　　75, 101, 122
sustainability gap analysis 45
Sustainability Life Cycle Assessment
　　(SLCA) 68
sustainable development 2–4, 14–32, 37,
　　39, 42–53, 61, *62,* 63, 65, 70, 75,
　　80–86, 89, 91, 94, 96, 97, 100, 108,
　　114–117, 119, 122, 123, 126–127
　PVC industry in global regions 83–86
Sustainable Development Goals (SDGs) 3, 20,
　　29, 60–63, *62, 63,* 80
Sweden 25, 41, 44
Swedish 25
synthetic plastics 12
synthetic substances 2, 53, 99
Syrian civil war 37

System Conditions (SCs) 3, 26, *26,* **27,** 46, 49,
　　63, 68, *68,* 69, 70, **71–74,** 83, 96,
　　100, 101
systemic screening 108
SYSTEMIQ 20
systems-based ASF tool 118
systems science model 26, *26,* 63
systems scientist 126

Tamoxifen 17
tar deposit(s) 11
taxol 17
tea 17
Teflon (polytetrafluoroethylene) (PTFE) 108
termite(s) 6, 7
Thatcher, Margaret 113
The Natural Step (TNS) 3, 21, 24–26, *26,* 27,
　　27, *27,* 28, 45, 46, 49, 63, *64,* 68,
　　68, 69, 70, *71–74,* 83, 91, 92, 96,
　　100, 101, 118, 120
　organisation's website **28**
　potential sustainability issues **71–74**
　Sustainability Challenges for PVC 46, *46,*
　　49, **50–51,** 59–60, **61,** *64,* 67, 70,
　　75, 76–79, 90–96, 127
thermodynamics 25, 83
thermoforming 36
timber 21, 24, 92–93, 101
tin 9, 10
titanium dioxide **50**
tobacco 22
'tomorrow's problem(s)' 22, 42
*Towards a Non-toxic Environment
　　Strategy* 120
toxicity 38, 39, 95
toxin-free environment 119
trade association(s) 58, 85, 86
trade-off(s) 28, 117
'traffic lights' colour coding 69–70, 97,
　　98–99
transportation 11, 37, 47
"Treading lighter on the Earth" 116
Twenty years of the Polyvinyl Chloride (PVC)
　　sustainability challenges 91

ultrasound 12
UNEP *see* United Nations Environment
　　Programme (UNEP)
unequal access 19
United Kingdom (UK) 2, 20, 41–43, 45, 48,
　　54, 58, 80, 82, 86, 91, 113, 123

United Nations Conference on Environment
   and Development 114
United Nations Environment Programme
   (UNEP) 23
'unsustainability hotspots' 69
upper atmosphere 38
urban development 113
US 20, 23, 41, 86, 90, 95,
   108, 113
US Clean Air Act (1970) 23

value chain 2–4, 24, 43, 46, 48, 49, 81, 83–85,
   90, 91, 93, 94, 115, 126
Vantage Vinyl 83
vehicle fuel 21, 39
vested interest(s) 22, 118
vinum 34
Vinyl2010 3, 58, 59, *59*, 91
vinyl (s) 11, 34, 37, 54, 75, 82–85, 97
vinyl chloride monomer (VCM) 33–34, *34*,
   38, 41, 44, 90
Vinyl Council of Australia 84
Vinyl Institute of Canada 84
VinylPlus 3, *59*, 59–60, *60*, **61**, 62, 63–67, *64*,
   67, 70, *76–79*, 81, 82, 84, 85, 91,
   92, 96, 118, 120
   accreditation schemes 64–67, **65**
   ASF approach 67–70, *68*
   programmes 59–60, *60*
   progress report 70, **76–79**
   Sustainable Development Goals (SDGs)
      60–63, *62, 63*
   TNS Sustainability Challenges for PVC
      60, **61**
VinylPlus Annual Reports 60
VinylPlus Pathways 3, 63–64, *64*, 67,
   76–79, 83

VinylPlus Product Label 65, 66, *66*, 67, 82
   criteria for accreditation 65, **65**
VinylPlus Progress Report 2022 81
VinylPlus Progress Report 2023 81
VinylPlus Supplier Certificate (VSC) 66, 67,
   67, 69, 82
vinyl record 11, 37, 97
Vinyl Sustainability Council 83
virgin PVC 39, 75, 120
vision 24, 27, 45, 62, 80, 117, 118,
   120–122, 127
Voltaire 55, 118, 119
voluntary commitment 2, 3, 41, 47, 59, **61**, 65,
   81, 85, 90, 91, 95, 115, 123

wastage 21, 119, 121, 122
waste 1, 6, 19, 23, 24, 26, 29, 38–40, 44, 47,
   53, 54, 59, 75, 81, 82, 85, 112,
   118–121, 123
   management of 20
water pipes 36, 62, 93
wildlife 22, 39
window profile(s) 24, 37, 39, 44, 69, *71–74*,
   92, 93, 100, 101
wine 17, 34
wipe-clean surface (s) 37
wire(s) 34, 37, 94
wood 11, 12, 24, 93
wood ants 6
World Trade Organization (WTO) 122
Worldwide Fund (WWF) 92

X-ray 12

yew tree(s) 17

zoonotic diseases 18